Energy Efficient Buildings
The Concept of Zero Carbon Emissions Green Building

Ioannis S. Vourdoubas

ELIVA PRESS

ELIVA PRESS

Ioannis S. Vourdoubas

Climate change is a severe global environmental problem threatening the sustainability of ecosystems worldwide and causing major catastrophes. Mitigation of climate change requires the sharp decrease of GHG emissions into the atmosphere and many countries are targeting to become carbon neutral by 2050. Buildings utilize large amounts of energy while in Europe they consume 40% of the overall energy use. European directives and legislation in member states promote the increase in energy efficiency and performance of public and private buildings transforming them to "nearly zero energy buildings". These green buildings have lower energy consumption, utilize less fossil fuels, more renewable energies while they emit less GHGs. A slightly different concept to "nearly zero energy building" is the concept of green "zero carbon emissions building". These types of energy efficient and green buildings utilize renewable energies to generate the required heat zeroing their carbon emissions. They also generate "green electricity" for their own consumption while they offset annually the grid electricity use. Current advances in various benign renewable energy technologies allow their use for heat, cooling and electricity generation in buildings. Many of these technologies are mature, reliable and cost-effective. They can be used for green energy generation either on-site or off-site while they can also recharge the electric car's batteries owned by the building's residents. The book consists of two parts. The first part includes six papers related with the creation of zero carbon emissions green residential buildings. The second part includes eight papers related with the creation of other types of energy efficient ZCEBs including hospitals, schools, offices, academic institutions, prisons, museums and swimming pools. The book could be useful to architects, engineers, building construction companies and companies manufacturing and installing various renewable energy systems. It could be also useful to public authorities interesting to improve the energy performance in public buildings as well as to policy makers who want to decrease the carbon emissions in the building sector complying with the global efforts for climate change mitigation.

Published by Eliva Press SRL
Address: MD-2060, bd.Cuza-Voda, 1/4, of. 21 Chişinău, Republica Moldova
Email: info@elivapress.com
Website: www.elivapress.com

ISBN: 978-1-63648-098-5

© Eliva Press SRL, 2021
© Ioannis S. Vourdoubas
Cover Design: Eliva Press SRL

No part of this book may be reproduced or utilized in any form or by any means, electronic or mechanical, including photocopying, recording, or by any information storage and retrieval system, without permission in writing from Eliva Press.

All rights reserved.

PREFACE

Buildings consume a large amount of overall energy consumption while in Europe their share to total energy consumption is approximately at 40%. The urgent necessity to cope with the most important global environmental problem in our era - Climate Change – requires the sharp decrease of carbon emissions into the atmosphere and the replacement of fossil fuels with benign, carbon free, energy sources. Reduction of energy consumption in the building sector is very important for the EU strategy to meet the target of climate change mitigation. European Union currently promotes the improvement of energy performance in buildings lowering their energy consumption, fossil fuels use and increasing the use of renewable energy sources (RES) in them. According to European legislation existing public and private buildings should be refurbished to become _Nearly Zero Energy Buildings (NZEBs)_ while new constructed buildings should follow the same principle. A NZEB is conceived as a building that has:

a) Very low energy consumption,
b) Minor use of fossil fuels in covering its requirements in heat, cooling and electricity, and
c) Major use of renewable energies, preferably on-site, for covering its energy requirements.

The concept of energy efficient green NZEBs has different interpretations and various definitions have been proposed. Current debates regarding this definition are related with the distinction between energy and primary energy use, metrics and energy conversion factors, use of RES on-site, nearby or far away, etc. Transformation of the EU's building stock in green NZEBs will result in many environmental, economic, social and geo-political benefits including:

1. Lower fossil fuels consumption,
2. Less CO_2 emissions into the atmosphere,
3. Increasing use of RES in covering the energy needs,
4. Increase of energy security,
5. Increase of energy self-sufficiency and lower dependence on fossil fuels resources of other countries,
6. Less fossil fuels imports from non-EU countries,
7. Assistance to EU industry related with building construction, energy saving and renewable energy systems construction and installation,
8. New jobs creation in the building and energy sector,
9. Increase of energy democracy allowing citizens to generate the energy that they consume it their residential buildings, including the electricity required

for re-charging the batteries of their electric vehicles, becoming energy pro-cumers,
10. Promotion of research and development in EU as well as promotion of scientific and technological innovations in the building and energy sectors.

Current work is related with a slightly different approach to NZEBs. This new concept is the "*Net zero carbon emission building*" (*NZCEB*). This type of energy efficient green building can be conceived as a grid-connected or autonomous building which does not emit CO_2 into the atmosphere due to operating energy use. A NZCEB:

a) Does not use fossil fuels for covering its energy requirements. Its energy needs are covered with the use of RES, and

b) If the building uses grid electricity generated by fossil fuels it should be offset with green electricity derived by RES including solar-PV electricity.

It should be noted that creation of NZCEBs requires that existing legislation should allow the offsetting of grid electricity use in buildings with green electricity. Currently in EU countries net-metering regulations allow the offsetting of grid electricity consumption in buildings with green electricity, generated preferably on-site, like solar electricity. It should be mentioned that many renewable energy technologies which could be used for energy generation in buildings are currently mature, reliable and cost-effective. These include among others:

1. *Solar energy for heat and electricity generation,*
2. *Solid biomass for heat generation, and*
3. *Low enthalpy geothermal energy combined with high efficiency heat pumps for heat and cooling generation.*

The book is consisted of fourteen papers which have been published during the recent years in various international peer-reviewed Scientific Journals. They present the methodology for creating green NZCEBs with examples of case studies for different types of buildings. The case studies are mainly focused in the island of Crete, Greece and the use of indigenous RES. However, the same methodology can be used in other countries with different availability of RES.

The book is separated in two parts. The first part includes six (6) papers related with the creation of net zero carbon emissions green residential buildings. The possibility for using RES for re-charging the batteries of the electric cars owned by the residents is mentioned. The second part includes eight (8) papers related with the creation of other types of NZCEBs including hospitals, schools, offices, academic institutions, prisons, museums and swimming pools. The methodology followed is focused in operating energy consumption in buildings. However, the possibility of zeroing carbon emissions due to embodied energy which includes the energy consumed during their construction, refurbishment and demolition is also stated. In the case studies presented there is not any distinction regarding the installation of renewable energy systems on-site, nearby or far away the building.

The book indicates the way to create new or refurbished green buildings with neutral climate footprint due to energy use. This can be achieved, in the different case studies presented, taking into account the availability of various RES in Greece. However, in other territories there are probably the same as well as other indigenous RES which could be used for "green energy" generation in buildings achieving the same goal. The concept of green NZCEBs has been promoted through the EU funded project "Promotion of near zero CO_2 emission buildings due to energy use" financed in the framework of INTERREG EUROPE and implemented during the period 2016-2020. The methodology proposed for achieving NZCEBs has the advantages that:

a) *Buildings do not utilize fossil fuels neither emit CO_2 into the atmosphere due to energy use, and*
b) *They cover all their energy requirements with RES,*
c) *This can be achieved in a cost effective way, and*
d) *It complies with the EU goal to become carbon neutral continent by 2050.*

The conclusions drawn indicate that creation of new or refurbished green *"net zero carbon emission buildings"* is technically and economically feasible. Various mature, reliable and cost-effective renewable energy technologies can be used for covering all their energy needs in heat, cooling and electricity.

The book could be useful to architects, engineers, building construction companies and companies constructing and installing various renewable energy systems. It could be also useful to anyone who wants to zero its carbon footprint due to operating energy use in his residential building as well as due to electricity required for recharging the batteries of his electric car. Finally, it could be useful for public authorities trying to improve the energy efficiency and sustainability in public buildings as well as to local, regional and national policy makers who want to develop appropriate policies for reduction of carbon emissions in the building sector and to mitigate climate change.

December, 2020
Crete, Greece

Contents

PART 1 13
[1] Creation of Zero CO_2 Emissions Residential Buildings due to Energy Use. A Case Study in Crete, Greece 13
1. Introduction 13
2. Literature survey 13
3. Energy Supply for Buildings in Crete, Greece: Current Situation 16
3.1 Energy Use in Residential Buildings 16
3.2 Life Cycle Energy Use in Residential Buildings 17
3.3 Use of Renewable Energies in Residential Buildings 17
3.4 Zeroing CO_2 Emissions due to Energy Use in Residential Buildings 18
4. Design of Zero CO_2 Emission Houses due to Energy Use 19
4.1 Design of Zero CO_2 Emission Houses due to Energy Use in Crete, Greece 19
4.2 Use of Solar Energy and Solid Biomass in a house in order to zero its CO_2 emissions 20
4.3 Use of Solar Energy and Geothermal Energy in a house in order to zero its CO_2 emissions 21
4.4 Cost Analysis 22
4.5 CO_2 Emission Savings 23
5. Conclusions 24
[2] Realization of a small residential building with net zero CO_2 emissions due to energy use in Crete, Greece 27
1. Introduction 27
2. Literature survey 27
2.1 Energy use in residential buildings 27
2.2 Life cycle energy analysis in buildings 28
2.3 Use of renewable energies in residential buildings 29
2.4 Zero energy buildings 30
2.5 Zero CO_2 emission buildings due to energy use 32
3. Differences between near zero energy buildings and near zero CO_2 emission buildings 33
4. Realization of a small residential building with zero CO_2 emissions due to energy use located in Crete, Greece 34
4.1 Building characteristics 35

4.2 Requirements for zeroing the carbon footprint in residential buildings 35
4.3 Energy loads 35
4.4 Use of renewable energy technologies 36
4.5 Energy balance in buildings 37
5. Cost analysis 37
6. Environmental analysis 38
7. Discussion 39
8. Conclusions 40
References 42
[3] Creation of zero CO_2 emissions residential buildings due to operating and embodied energy use in the island of Crete, Greece 45
1. Introduction 45
2. Literature survey 45
2.1 Low energy buildings 45
2.2 Environmental impacts of buildings due to energy use 46
2.3 Energy consumption in buildings 47
3. Operating and embodied energy in buildings 49
4. Zero CO_2 emission residential buildings due to operating energy use 50
5. Zero CO_2 emission residential buildings due to operating and embodied energy use 51
5.1 Mathematical formulation 52
5.2 Adjustment due to grid electricity generation from renewable energies and nuclear power 53
5.3 Estimation of the required solar-PV system when solid biomass is used for space heating 53
5.4 Estimation of the required solar-PV system when a high efficiency heat pump is used for space heating 54
6. Discussion 57
7. Conclusions 58
References 59
[4] Use of renewable energies for creation of net zero carbon emissions residential buildings in northern Greece 62
1. Introduction 62
2. Literature survey 62
2.1 Use of solid biomass in individual heating systems 62
2.2 Use of solid biomass in district heating systems 63
2.3 Use of waste heat in district heating systems 63

2.4 Use of solar thermal energy for domestic hot water production 64
2.5 Use of solar-PV systems 64
2.6 Net zero energy and net zero carbon emission buildings 65
3. Energy use in residential buildings 66
4. Requirements for net zero carbon emission buildings due to energy use 67
5. Availability of solar energy, solid biomass and waste heat in northern Greece 67
6. Use of solid biomass and solar energy for covering all the energy needs in residential buildings 69
7. Use of waste heat and solar energy for covering all the energy needs in residential buildings 71
8. Discussion 73
9. Conclusions 73
References 74
[5] Review of sustainable energy technologies use in buildings in Mediterranean basin 77
1. Introduction 77
2. Literature survey 77
2.1 Solar thermal systems with flat plate collectors 77
2.2 Solar parabolic collectors 78
2.3 Solar thermal cooling 78
2.4 Solar photovoltaic use 79
2.5 Wind turbine use 79
2.6 Solid biomass use 80
2.7 High efficiency heat pumps 80
2.8 Co-generation of heat and power 81
2.9 District heating using biomass 81
2.10 District heating using waste heat 81
2.11 Creation of net zero energy buildings with net zero carbon emissions due to energy use 82
3. Use of sustainable energy technologies for energy generation in buildings 82
4. Assessment of various sustainable energy technologies use in buildings 85
5. Creation of net zero energy buildings in Mediterranean basin 88
6. Discussion 89
7. Conclusions 90
References 91
[6] Creation of net zero carbon emissions residential buildings due to energy use in Mediterranean region. Are they feasible? 95

1. Introduction 95
2. Literature survey 95
2.1 Energy consumption in buildings 95
2.2 Embodied and operating energy consumption 96
2.3 Net zero energy buildings 97
2.4 Net zero carbon emission buildings 98
2.5 Use of solar electricity to charge batteries in electric cars 99
3. Energy requirements in residential buildings 100
3.1 Operating energy use 100
3.2 Embodied energy use 102
3.3 Energy required for recharging batteries of electric vehicles 102
3.4 Nearly zero energy residential buildings 102
4. Use of renewable energy technologies in residential buildings 104
4.1 Solar thermal energy 104
4.2 Solar photovoltaic energy 104
4.3 Solid biomass 104
4.4 Low enthalpy geothermal energy 104
4.5 Other sustainable energy technologies 105
5. Net zero carbon emission buildings 105
6. Compensation of grid electricity consumption in residential buildings with green electricity generated in them 106
7. A case study of a residential building with net zero carbon emissions due to energy use located in Greece 106
7.1 Solar electricity generation covering all its operating energy needs 107
7.2 Solar electricity generation covering its embodied energy 108
7.3 Solar electricity generation for recharging the electric batteries of two vehicles. 109
8. Economic and Environmental considerations 109
8.1 Cost estimations 109
8.2 Environmental considerations 109
9. Discussion 110
10. Conclusions 111
References 112

PART 2 116

[7] Creation of net zero CO_2 Emissions Hospitals Due to Energy Use. A Case Study in Crete, Greece 116
1. Introduction 116
2. Literature survey 116
3. Energy consumption in hospitals 119
4. Use of renewable energies in hospitals 120
5. Creation of hospitals with zero CO_2 emissions due to energy use 121
5.1 Use of solar energy and solid biomass for covering all the energy needs in a hospital in Crete, Greece.. 121
6. Economic Considerations 125
7. Environmental benefits due to renewable energies use in a hospital 126
8. Conclusions 127
References 127
[8] Creation of zero CO_2 emissions school buildings due to energy use in Crete, Greece... 130
1. Introduction 130
2. Literature survey 130
3. Energy consumption in school buildings in Greece................................ 133
4. Use of renewable energies for energy generation in school buildings in Crete, Greece... 134
5. Design of school buildings with zero CO_2 emissions due to energy use in Crete, Greece... 135
5.1 Description of a school building which covers all its energy requirements with solar electricity and solid biomass in Crete, Greece 136
5.2 Description of a school building which covers all its energy requirements with solar electricity and a geothermal heat pump in Crete, Greece 136
6. Economic considerations 137
7. Environmental and social considerations 138
8. Conclusions 139
References 140
[9] Possibilities of creating swimming pools with zero CO_2 emissions due to energy use. A case study in Crete, Greece 142
1. Introduction 142
2. Literature review.. 142
3. Use of renewable energies in swimming pools...................................... 144
4. Creation of zero CO_2 emission swimming pools due to energy use 147

5. Creation of a zero CO_2 emissions swimming pool. A case study in Crete, Greece. 147
5.1 Use of solar thermal energy and solid biomass for heat generation 148
5.2 Use of solar thermal energy and a ground source heat pump for heat generation 149
5.3 Comparison of two different combinations of renewable energy systems providing all the required energy in the pool 150
6. Conclusions 151
References 152
[10] Creation of zero CO_2 emissions office buildings due to energy use. A case study in Crete, Greece.. 155
1. Introduction 155
2. Literature survey 155
3. Availability of renewable energy sources in the island of Crete, Greece 159
4. Creation of zero CO_2 emissions office buildings.. 161
5. Design of an office building with zero CO_2 emissions in Crete, Greece.......... 161
5.1 Use of solar energy and solid biomass for covering all its energy requirements 162
5.2 Use of solar energy and low enthalpy geothermal energy for covering all its energy requirements 163
6. Discussion and Conclusions... 164
References 165
[11] Energy consumption and carbon emissions in Venizelio hospital in Crete, Greece. Can it become carbon neutral? 168
1. Introduction 168
2. Literature survey 168
2.1 Energy consumption in hospitals 168
2.2 Use of renewable energies in hospitals 170
3. Description of the hospital... 171
4. Energy use and carbon emissions in hospitals 173
5. Use of various renewable energy technologies for covering the energy requirements in Venizelio hospital 174
5.1 Use of solar thermal energy.. 174
5.2 Use of solid biomass... 174
5.3 Use of low enthalpy geothermal energy with heat pumps........................... 175
5.4 Use of solar-PV energy... 175
6. Cost estimations.. 176

7. Discussion 178
8. Conclusions 178
References 179
[12] Energy consumption and carbon emissions in an Academic Institution in Greece. Can it become carbon neutral? 183
1. Introduction 183
2. Literature survey 183
3. Description of the Academic Institute 188
4. Energy consumption in the Institute 188
5. Carbon emissions due to operating energy use in the Institute 189
6. Use of renewable energy technologies for energy generation in the Institute 190
7. Requirements for a carbon neutral Institution 190
8. Sizing various sustainable energy systems covering all energy requirements 191
8.1 Sustainable energy technologies use 191
8.2 Estimation of energy loads 191
8.3 Sizing the necessary sustainable energy systems 192
8.4 Capital cost of sustainable energy systems 193
9. Discussion and conclusions 193
References 194
[13] Possibilities of creating net zero carbon emissions cultural buildings. A case study of the museum in Eleftherna, Crete, Greece 197
1. Introduction 197
2. Literature survey 197
2.1 Energy consumption in museums and historical buildings 197
2.2 Use of renewable energies in museums 199
2.3 Zero carbon emissions buildings 201
3. Requirements for zeroing net carbon emissions due to energy use 202
4. The Archaeological Museum of Eleftherna in Crete, Greece 203
5. Use of renewable energy technologies in the museum 203
5.1 Use of solar-PV energy for electricity generation 204
5.2 Use of high efficiency heat pumps for air-conditioning 204
6. Environmental and economic considerations 204
7. Discussion 205
8. Conclusions 206
References 207

[14] Possibilities of creating net zero carbon emission prisons in the island of Crete, Greece. 210
1. Introduction 210
2. Literature survey 210
2.1 Energy consumption in prisons 210
2.2 Use of sustainable energies in prisons 212
2.3 Carbon sequestration from tree plantations 213
3. Energy consumption in prisons 215
3.1 Operating energy 215
3.2 Embodied energy 215
4. Possibilities of using renewable energy technologies in Cretan prisons 216
5. Requirements for net zero carbon emissions prisons 217
6. A case study of a net zero carbon emissions prison located in Crete, Greece 217
6.1 Sizing the solar energy systems generating all energy requirements 217
6.2 Sizing the solar energy systems and the area of the tree plantation for zeroing carbon emissions 219
7. Opportunities and barriers in "greening" Cretan prisons 220
8. Discussion 221
9. Conclusions 221
References 222
[15] Papers included in the book 225
Part 1 225
Part 2 225

PART 1

[1] Creation of Zero CO_2 Emissions Residential Buildings due to Energy Use. A Case Study in Crete, Greece

1. Introduction

Energy consumption is related with severe global environmental problems, including climate change, resulting in undesired economic and social consequences. European buildings are responsible for approximately 40% of the total energy consumption and many efforts are focused either in reducing their total energy use or in replacing their use in fossil fuels with renewable energies. The legal framework in Europe, after 2000, aims in improving energy behavior of buildings and reducing their CO_2 emissions due to energy use categorizing them according to their energy performance. A new E.U. directive issued in 2010 aims at creating new buildings with very low energy consumption after 2020. Priority is given to public buildings which must be transformed in near zero energy buildings from 1/1/2019. Progress in energy saving techniques and technologies as well as current developments in various renewable energy technologies for heat, cooling and power generation have increased their reliability and allow their use in buildings in a cost effective way. At the same time new financial tools and governmental incentives promote the use of sustainable energy technologies in buildings. Energy is consumed in various stages over the life span of the building during its initial construction, during its refurbishments, during its operation and finally during its demolition. Construction and operation of grid connected residential buildings with zero CO_2 emissions due to energy use have not been reported so far in Greece. However, recent advances in various renewable energy technologies allow their combined use in a cost effective way in residential buildings located in areas with mild climate like in Crete, Greece in order to zero their CO_2 emissions and their carbon footprint due to energy use in them.

2. Literature survey

Estimation of energy consumption and CO_2 emission due to housing construction in Japan has been investigated [1]. Energy consumption at 3-10 GJ/m^2 (833-2,777 KWh/m^2) has been found depending on the type and construction of the building while its CO_2 emissions during the construction stage varied between 250-850 kg/m². The impact of urban climate on the energy consumption of buildings has been presented [2]. It has been found that when the mean heat island intensity exceeds 10 °C during the summer in the city of Athens the cooling load of urban buildings may be doubled and the peak electricity load for cooling purposes may be tripled. During the winter period the heating load of centrally located urban buildings is found to be reduced by 30%. The progress towards sustainable energy buildings has been studied [3]. Increase of energy efficiency and use of renewable energies are necessary in modern buildings. An energy performance assessment of existing dwellings has been proposed [4]. According to the authors E.U. final energy consumption in 2002 in the building sector corresponds at 40.3% of the total EU-25 final energy use. The consumed energy in dwellings corresponds at 25.4% of the total final energy use. A methodology for life cycle building's energy rating has been developed [5]. Apart from the operating energy consumed during its operation the embodied energy is usually not taken into account. In buildings with "net zero energy consumption" the only life-cycle energy use is the embodied energy. In a case study of a detached house in Ireland, the authors have estimated the embodied energy at 1,000 KWh/m². A review of lessons learnt from natural world for sustainable buildings solutions has been published [6]. According to the authors one basic pillar of buildings sustainability is the energy efficiency and the reduction of greenhouse gas emissions. An empirical assessment of the Hellenic building stock, its energy consumption as well as its carbon emissions has been published [7]. The annual total energy consumption in residential buildings in EU averaged at 150-230 KWh/m² in 1990s, which is distributed for space heating 70%, water heating 14%, and operation of various electrical appliances and lighting 12%. The residential annual energy use per capita among European countries varies from 1,500-5,000 KWh/person in Southern Europe to over 8,000 KWh/person in northern Europe. A comprehensive energy and economic analysis of a zero energy house versus a conventional house has been published [8]. The authors have used, in zero energy houses, various energy saving techniques, solar-PV panels and a solar thermal water heater proving that some of them were economically cost-effective. A report on energy consumption of residential buildings has been published [9]. The authors have used a neural network approach to model and estimate energy consumption time series for a residential building in Athens. These models gave accurate predictions of future energy consumption in buildings. Heating energy consumption and resulting environmental impacts of European apartment buildings has been presented [10]. According to the authors the average annual heating energy consumption was at 174.3 KWh/m². A study concerning energy savings in Danish residential buildings has been published [11]. The authors concluded that there are many opportunities in order to realize various energy saving measures which are also cost-effective. A review on building's energy consumption has been published [12]. Energy consumption by end uses varies widely among various countries while the largest share of energy consumption corresponds to air-conditioning. The perspective of very low energy homes in USA has been presented [13]. The author reported that buildings with very low energy use can be easily achieved in North America. He considers that energy saving is a prerequisite for installation of solar water heaters and solar electricity in order to obtain a near zero energy home. A case study of zero energy house design in UK has been presented [14]. The authors considered a house with low energy consumption which used solar energy for hot water production, a solar-PV and a wind turbine for electricity generation. They suggested a three-step process for designing a zero energy house. Firstly, analysis of a local climate data, secondly, application of passive design methods to minimize energy requirements and thirdly use of renewable energy systems to cover all the energy needs. An analysis of life cycle zero energy buildings has been presented [15]. Apart from the operating energy which is used during the operation of the house its embodied energy which is used during its construction, refurbishment and demolition should be taken into account. A review of life cycle energy analysis of buildings has been presented [16]. The authors have considered that the annual life cycle energy (primary) requirements of conventional residential buildings fall in the range 150-400 KW/m² which includes both operating and embodied energy. The operating energy corresponds at 80-90% of the total energy use while the embodied energy at 10-20%. The authors reported that demolition energy had a negligible effect to the total energy balance of the building. A review of zero energy buildings has been presented [17]. The authors considered that zero energy buildings is the future target for the design of buildings and this requires a clear and consistent definition of them as well as a commonly agreed calculation methodology. A study on the influence of occupant's behavior on building's energy consumption has been presented [18]. The authors investigated the building occupant's behavior and activities as well as their social and economic status on energy consumption. A feasibility study of a zero energy home in Newfoundland has been presented [19]. The authors have concluded that a 10 KW wind turbine can cover all the energy needs of a grid connected residential house in Newfoundland. A critical look at zero energy buildings has been presented [20]. The authors have defined four types of zero energy buildings as follows: (a) Net zero site energy; (b) Net zero source energy; (c) Net zero energy cost and (d) Net zero energy emissions while they have commented on the advantages and disadvantages of each one of them.

3. Energy Supply for Buildings in Crete, Greece: Current Situation

3.1 Energy Use in Residential Buildings

Energy is used in residential buildings in various operations including space heating and cooling, hot water production, lighting and operation of various electric appliances. Energy consumption in residential buildings depends on many factors including local climate, type and quality of construction of the building, social and economic status of the occupants, type and quality of equipment used etc. Monthly air temperatures in Chania, Crete, Greece vary between 7.9-14.1 °C in January and 21.2-30.6 °C in July. Solar irradiance in Chania, Crete, Greece at tilt 30 degrees varies between 83 KWh/m² in January to 208 KWh/m² in July with monthly average 145 KWh/m². The highest percentage of energy used in residential buildings corresponds to space heating and it varies approximately at 60-70% to total energy consumption. Typical energy consumption of a residential building in Crete, Greece is presented in Table 1.

Table 1. Typical energy consumption in a residential building located in Crete, Greece [1]

Sector	% of energy used	Annual energy consumption (KWh/m²)
Space heating	63	107.1
Hot water production	9	15.3
Lighting	12	20.4
Operation of various appliances including space cooling	16	27.2
Total	100	170

[1] Source: Own estimations

3.2 Life Cycle Energy Use in Residential Buildings

Energy is used in residential buildings during their construction, refurbishment, operation and demolition. Considering that the life of a residential building is 50 years energy is used:

During its initial construction,
During its refurbishment and maintenance,
During its operation, and
During its demolition

Total energy use in a residential building during its life cycle is the sum of the abovementioned energies used in various time intervals. Energy consumed during its construction, refurbishment and demolition is considered as embodied energy in the building. In the case of using renewable energy systems in the building providing heat or/and electricity a life cycle analysis of the energy systems used should be made in order to have an accurate calculation of the net energy generation in the residential building.

3.3 Use of Renewable Energies in Residential Buildings

Various renewable energies can be used in residential buildings for generation of heat, cooling and electricity. They include: solar thermal and solar-PV energy, wind electricity, solid biomass and geothermal energy. Passive solar energy techniques can be used in order to lower the heating and cooling loads in the building resulting in less energy use. Passive solar architecture is currently applied in various residential and non-residential buildings. Solar thermal energy is extensively used in Greece for domestic hot water production (DHW) in residential buildings and hotels. Usually simple thermosiphonic systems are used and due to high irradiance in Greece these systems can cover the needs for DHW during most days of the year. Each m^2 of flat plate collector is equivalent to 0.7 KW$_{th}$ and produces annually in Crete approximately 860 KWh/KW. Solar photovoltaic energy is also used in grid connected and autonomous residential buildings. During recent years decrease in prices of solar-PV cells combined with attractive governmental incentives concerning their installation have increased their use in buildings. Poly-crystaline, mono-clystaline and thin film solar cells are used generating in Crete annually approximately 1,500 KWh/KWp. Usually the photovoltaic cells are placed on the roofs of the buildings and the generated electricity is either injected into the grid or self-consumed. In autonomous residential buildings they can cover (combined with electric batteries) their electricity needs. Wind power can be used for electricity generation in residential buildings. In an autonomous house, if the mean annual air velocity is satisfactory, a wind turbine combined with an electric battery can cover part or all of its electricity needs preferably combined with solar cells. Solid biomass can be used for heat generation in residential buildings. Various types of solid biomass with heating values in the range of 3,700-4,200 Kcal/kg and various heating systems are used in order to cover all the heating needs of residential buildings including space heating and production of DHW. Biomass fueled heating systems include high efficiency fireplaces, wood stoves and central heating systems while their efficiencies vary between 0.7-0.8. In rural communities where the cost of solid biomass is low compared with the cost of natural gas, heating oil and electricity these heating systems are very popular in residential buildings. Low enthalpy geothermal energy with high efficiency heat pumps, particularly ground source heat pumps, can be used for heat and cooling generation in residential buildings. Heat pumps are very efficient energy devices having high coefficients of performance in the order of 3-4. They utilize electricity and can produce heat, cooling and hot water covering all the heating and cooling needs in residential buildings. They absorb or reject heat from the ground in low depth where the temperature remains almost constant all over the year resulting in their high efficiency.

3.4 Zeroing CO_2 Emissions due to Energy Use in Residential Buildings

Current European policies aim to minimize net energy use as well as greenhouse gas emissions in buildings resulting to nearly zero energy buildings after 31/12/2020 and particularly for public buildings after 31/12/2018 [21, 22]. In order to zero CO_2 emissions due to energy use in grid connected residential buildings two requirements must be fulfilled as follows

a)   Use of renewable energies instead of fossil fuels for heat generation, and

b)   Use of renewable energies for electricity generation which will be injected into the grid and will offset the grid electricity consumed annually. In that case electricity generated by fossil fuels is offset by green electricity generated from renewable energies.

It has been assumed that all grid electricity has been generated with fossil fuels but in most cases part of grid electricity has been derived from renewable energies. Without such consideration CO_2 emissions from the building are overestimated. In order to zero total CO_2 emissions both from operating energy and embodied energy use, electricity generation from renewable energies must be higher than grid electricity consumed annually from the residential building. Analytical calculations considering the sum of embodied and operating energy over the life span of the building are required for zeroing CO_2 emissions due both to embodied and operating energy consumption. In order to minimize total energy consumption in residential buildings application of passive and active energy saving techniques and technologies are required. Many of these techniques are cost-effective and their use is necessary in order to increase the sustainability in residential buildings. In the case of energy autonomous off-grid residential buildings all the annual electricity requirements will be covered with autonomous renewable energy systems combined with electric batteries.

4. Design of Zero CO_2 Emission Houses due to Energy Use

4.1 Design of Zero CO_2 Emission Houses due to Energy Use in Crete, Greece

Design of two houses in Crete, Greece with equal annual energy consumption at 170 KWh/m² which have net zero CO_2 emissions due to energy use in Crete, Greece is presented. Both houses are similar regarding their design and they are grid connected but they use a combination of different renewable energy technologies in order to zero their net annual CO_2 emissions due to operating energy use. Both houses have a covered surface of 200 m² and they use solar thermal energy, solar-PV energy, solid biomass and low enthalpy geothermal energy for covering all their energy needs. The first house is using solar energy and solid biomass and the second house solar energy and low enthalpy geothermal energy with heat pumps as is shown below. Energy consumption of both houses is the same and is presented in Table 2.

Table 2. Energy consumption in a house with covered surface at 200 m² located in Crete, Greece [1]

Sector	% of energy used	Specific annual energy consumption (KWh/m²)	Annual energy use (KWh)
Space heating	63	107.1	21,420
Hot water production	9	15.3	3,060
Lighting	12	20.4	4,080

Operation of various electric appliances including space cooling	16	27.2	5,440
Total	100	170	34,000

[1] *Source: Own estimations*

Solar irradiance is high in Crete-Greece and simple solar themosiphonic systems which are very popular in the island can produce hot water to cover approximately 80% of the annual needs in DHW. Solar-PV systems can generate annually in Crete approximately 1,500 KWh/KWp and the government allows with the net-metering initiative the installation of solar-PV cells in grid connected residential buildings in order to offset their annual electricity consumption. Locally produced solid biomass is used extensively for space heating in residential buildings particularly during the last 5-10 years. The current economic crisis in the country and the high taxation of the conventional heating fuels favors its broad utilization. Simple air to air or/and air to water heat pumps are used in residential buildings for space heating and cooling. Ground source heat pumps are not often used in residential buildings in Greece although they have high coefficients of performance and they are more energy efficient compared with conventional heat pumps. It is expected that their use will be increased in the coming years.

4.2 Use of Solar Energy and Solid Biomass in a house in order to zero its CO_2 emissions

A combination of solar thermal energy, solar-PV energy and solid biomass can be used in order to cover all the annual energy needs of the abovementioned house in Crete, Greece resulting in zeroing its net CO_2 emissions due to operating energy use. Solar thermal energy with a simple thermosiphonic system can cover 80 % of the annual needs in DHW of the house (2,448 KWh/year). Solid biomass can be used in order to cover all its space heating needs and additionally the remaining 20 % of its needs in DHW (22,032 KWh/year). A solar-PV system can be used in order to generate the same amount of electricity that the house consumes annually from the grid. The solar-PV system will be connected with the grid and it will feed all the generated electricity which is not self-consumed into it at 9,520 KWh/year. Design considerations for solar thermal, solar-PV and solid biomass burning systems are presented in section 2.3. Energy generated from solar energy and solid biomass in order to cover all the energy requirements of a house in Crete, Greece is presented in Table 3.

4.3 Use of Solar Energy and Geothermal Energy in a house in order to zero its CO₂ Emissions

A combination of solar thermal energy, solar-PV energy and low enthalpy geothermal energy can be used in order to cover all the energy needs of the abovementioned grid connected house in Crete, Greece resulting in zeroing its CO_2 emissions due to operating energy use. Solar thermal energy with a simple thermosiphonic system can cover up to 80% of the annual needs in DHW of the house which corresponds to 2,448 KWh/year. Space heating and space cooling as well as the rest of the required DHW can be covered with a low enthalpy geothermal heat pump having a C.O.P. equal to 3.5. The required energy generation by the heat pump is 21,420 KWh/year for space heating, 612 KWh/year, for 20% of the annual needs of DHW, and 1,088 KWh/year for space cooling, totally 23,120 KWh/year. Since the heat pump has a C.O.P. 3.5 its annual electricity consumption will be 6,606 KWh/year. A solar-PV system can be used in order to generate annually the electricity loads of the building which are 6,606 KWh/year for the operation of the heat pump, 4,080 KWh/year for lighting and 4,352 KWh/year for the operation of various electric appliances excluding space cooling.

Table 3. Energy generated from solar energy and solid biomass in order to cover all the energy requirements of a house in Crete, Greece

Energy system	Annual Energy generation (KWh/year)	Fuel used (kg/year)	Power of the energy system
Solar thermal	2,448	0	2.8 KW$_{th}$
Solar-PV	9,520	0	6.35 KW$_{el}$
Solid biomass burning	22,032	6,409	22 KW$_{th}$
Total	34,000		

Heating value of solid biomass, 3,700 kcal/kg; Annual electricity generation from the solar-PV system, 1,500 KWh/KWp; Annual heat generation from the solar thermal system, 860 KWh/KW$_{th}$; Efficiency of biomass heating system, 0.8, Operating period of the heating system, 1,000 hours/year.

The total electricity used annually in the house is 15,038 KWh. Therefore, the solar-PV system must generate annually and feed into the grid 15,038 KWh while its nominal capacity will be 10.03 KWp. The energy generated from solar and geothermal energy in order to cover all the energy needs of the abovementioned grid-connected house in Crete, Greece is presented in Table 4. Comparing Tables 3 and 4 it can be seen that the total consumed energy in the second case is higher at 6,606 KWh/year compared with the first case. This additional amount is the electricity required for the operation of the heat pump which is generated from the solar-PV system.

4.4 Cost Analysis

The capital cost of renewable energy systems in two cases has been estimated as well as their operating cost related with the fuel used. Since solar and geothermal energy have zero cost the only cost element is the cost of solid biomass. The results are presented in table 5. It can be seen, in table 5, that the capital cost of renewable energy systems in the first house is significantly lower compared with the same cost in the second house. In the second house the energy operating cost is zero compared with the first house where the operating cost is 641 €/year due to the cost of the solid biomass.

Table 4. Energy generated from solar and geothermal energy in order to cover all the energy needs in a house located in Crete, Greece

Energy system	Annual energy generation (KWh/year)	Power of the energy system
Solar thermal	2,448	2.8 KW$_{th}$
Solar PV	15,038	10.03 KW$_{el}$
Low enthalpy geothermal heat pump [1]	23,120	9.25 KW$_{th}$
Total	40,606	

[1]Assuming that the heat pump operates 2,500 hours annually.

Table 5. Capital and operating costs of the renewable energy systems used in the two houses [1]

1st House

Energy system	Power (KW)	Capital Cost (€)	Operating cost (€/year)
Solar thermal	2.8	2,000	0
Solar-PV	6.35	7,620	0
Solid biomass	22	11,000	641

burning			
Total		20,620	641

2nd House

Energy system	Power (KW)	Capital cost (€)	Operating cost (€/year)
Solar thermal	2.8	2,000	0
Solar-PV	10.03	12,036	0
Geothermal heat pump	9.25	18,500	0 (electricity is provided by the solar-PV system)
Total		32,536	0

[1] *Cost of solar thermal system at 2.8 KW, 2,000 €; Cost of solar-PV, 1,200 €/KW$_p$; Cost of geothermal heat pump 2,000 €/KW; Cost of solid biomass burning system 500 €/KW$_{th}$; Cost of solid biomass 0.1 €/kg; Maintenance and depreciation costs of the energy systems have not been taken into account.*

The capital cost of the abovementioned energy systems is approximately less than 10-12% of the current construction cost of the two houses in Crete-Greece.

4.5 CO$_2$ Emission Savings

The use of renewable energies in order to cover all energy requirements in residential buildings results in CO_2 emissions savings due to energy use in them. Emissions savings can be calculated in the abovementioned houses from the fossil fuels which are used covering all their energy needs. Assuming that grid electricity generated from fossil fuels as well as heating oil are both used for covering their heating, cooling and electricity needs, CO_2 emissions savings due to use of renewable energies instead of fossil fuels are presented in Table 6. It can be seen that although energy corresponding to electricity use is lower than the energy corresponding to heating oil, CO_2 emissions savings due to electricity use are higher than emissions savings due to the use of heating oil.

Table 6. CO_2 emissions savings due to use of renewable energies for covering all energy needs in a house located in Crete, Greece [1]

Fuel	Use	Energy content (KWh/year)	Annual CO_2 emissions savings (KgCO_2)	Specific annual CO_2 emissions savings (KgCO_2/m^2)
Electricity consumption	Lighting, space cooling and operation of various electric devices	9,520	9,415	47.07
Heating oil consumption	Space heating and DHW production	24,480	7,496	37.48
Total		34,000	16,911	84.55

[1]Electricity emission coefficient, 0.989 kgCO_2/KWh, Heating oil emission coefficient, 3.2 kgCO_2/kg, Net heating value of heating oil, 9,000 Kcal/kg.

5. Conclusions

Residential buildings consume large amounts of energy for covering their energy needs. In order to improve their sustainability, they must decrease their energy consumption using various energy saving techniques and technologies and replace fossil fuels with renewable energies. Recent advances in renewable energy technologies have increased their maturity, reliability and cost effectiveness so they can replace all conventional energy sources used zeroing their CO_2 emissions due to energy use in them. Combined use of different renewable energies in two similar grid connected houses located in Crete, Greece has been presented in order to cover all their operating energy needs. Renewable energy technologies used are well known, reliable and cost effective. Some of them like solar water heating and solid biomass burning are extensively used to day in residential buildings in Crete, Greece. The first house used solar thermal energy, solar-PV and solid biomass for covering all its energy needs while the second solar thermal energy, solar-PV and geothermal energy with heat pumps for the same purpose. The cost of installing renewable energy systems in the two houses is less than 10-12% of their construction cost and the annual CO_2 savings are slightly less than 17 tons. The cost of installing renewable energy systems in the first house is approximately 30% smaller compared with the same cost in the second house which though has zero energy operating cost. In order to verify these estimations further experimental work with field tests and detailed energy studies is required. Additionally, a more detailed cost analysis should be made taking into account maintenance and depreciation costs of the proposed renewable energy systems.

References

[1] Suzuki, M., Oka, T. & Okada, K. (1995). "The estimation of energy consumption and CO_2 emission due to housing construction in Japan", *Energy and Buildings*, 22, pp. 165-169.

[2] Santamouris, M., Papanikolaou, N., Livada, I., Koronakis, I., Georgakis, C., Argiriou, A. et al. (2001). "On the impact of urban climate on the energy consumption of buildings", *Solar Energy,* 70, pp. 201-216.

[3] Chwieduk, D. (2003). "Towards sustainable energy buildings", *Applied Energy*, 76, pp. 211- 217.

[4] Poel, B., Cruchten, G.V. & Balaras, C.A. (2007). "Energy performance assessment of existing dwellings", *Energy and Buildings*, 39, pp. 393-403.

[5] Hernandez, P. & Kenny, P. (2011). "Development of a methodology for life cycle building energy ratings", *Energy Policy,* 39, pp. 3779-3788.

[6] Godfaurd, J., Clements-Croome, D. & Jeronimidis, G. (2005). "Sustainable building solution: A review of lessons from the natural world", *Building and Environment*, 40, pp. 319-328.

[7] Balaras, C.A., Gaglia, A.G., Georgopoulou, E., Mirasgedis, S., Sarafidis, Y. & Lalas, D.P. (2007). "European residential buildings and empirical assessment of the Hellenic building stock, energy consumption, emissions and potential energy savings", *Building and Environment*, 42, pp. 1298-1314.

[8] Zhu, L., Hurt, R., Correa, D. & Boehm, R. (2009). "Comprehensive energy and economic analyses on a zero energy house versus a conventional house", *Energy*, 34, pp. 1043-1053.

[9] Mihalakakou, G., Santamouris, M. & Tsangrassoulis, A. (2002). "On the energy consumption in residential buildings", *Energy and Buildings,* 34, pp. 727-736.

[10] Balaras, C.A., Droutsa, K., Daskalaki, E. & Kontoyiannidis, S. (2005). "Heating energy consumption and resulting environmental impact of European apartment buildings", *Energy and Buildings,* 37, pp. 429-442.

[11] Tommerup, H. & Svendsen, S. (2006). "Energy saving in Danish residential building stock", *Energy and Buildings*, 38, pp. 618-626.

[12] Perez-Lombard, L., Ortiz, J. & Pout, C. (2008). "A review on buildings energy consumption information", *Energy and Buildings,* 40, pp. 394-398.

[13] Parker, D.S. (2009). "Very low energy homes in the United States: Perspectives on performance from measured data", *Energy and Buildings,* 41, pp. 512-520.

[14] Wang, L., Gwilliam, J. & Jones, P. (2009). "Case study of zero energy house design in U.K.", *Energy and Buildings,* 41, pp. 1215-1222.

[15] Hernandez, P. & Kenny, P. (2010). "From net energy to zero energy buildings: Defining life cycle zero energy buildings", *Energy and Buildings,* 42, pp. 815-821.

[16] Ramesh, T., Prakash, R. & Shukla, K.K. (2010). "Life cycle energy analysis of buildings: An overview", *Energy and Buildings,* 42, pp. 1592-1600.

[17] Marszal, A.J., Heiselberg, P., Bourrelle, J.S., Mussal, E., Voss, K., Sartori,, I. et al. (2011). "Zero energy buildings–A review of definitions and calculation methodologies", *Energy and Buildings,* 43, pp. 971-979.

[18] Yu,, Z., Fung, B.C.M., Haghighat, F., Yoshino, H. & Morofsky, E. (2011). "A systematic procedure to study the influence of occupant behavior on building energy consumption", *Energy and Buildings,* 43, pp. 1409-1417.

[19] Iqbal, M.T. (2004). "A feasibility study of a zero energy home in Newfoundland", *Renewable Energy,* 29, pp. 277-289.

[20] Torcellini, P., Pless, S., Deru, M. & Crawley, D. (2006). " Zero energy buildings: A critical look at the definition", Conference paper NREL /CP-550-39833, *ACEEE summer study, Pacific Grove, California,* August 14-18, 2006.

[21] E.U. directive 2010/31 on the energy performance of buildings.

[22] E.U. directive 2002/91 on the energy performance of buildings.

[23] Gaglia, A.G., Balaras, C.A., Mirasgedis, S., Georgakopoulou, E., Sarafidis, Y. & Lalas, D.P. (2007). "Empirical assessment of the Hellenic non-residential building stock, energy consumption, emissions and potential energy savings", *Energy Conversion and Management,* 48, pp. 1160-1175.

[2] Realization of a small residential building with net zero CO_2 emissions due to energy use in Crete, Greece

1. Introduction

Climate change consists of the most severe environmental problem threatening the stability and existence of natural ecosystems and human societies worldwide. Its mitigation requires the significant decrease of GHG emissions in all sectors and the lower use of fossil fuels in energy generation. Buildings require large amounts of energy for covering their needs in heat, cooling and electricity. Decreasing their energy consumption and their fossil fuels use is important for reducing their carbon footprint. Use of renewable energies in residential buildings for heat, cooling and electricity generation can result in lowering or zeroing their carbon emissions due to energy use. Various renewable energy technologies are mature, reliable and cost-effective. They can be used in residential buildings covering part, or even all, of their energy requirements in heat, cooling and electricity. Current research is important because it indicates that due to progress and advances in benign green energy technologies their use in residential buildings for energy generation is feasible resulting in zeroing their net carbon footprint complying with the European and global efforts regarding climate change mitigation.

2. Literature survey

2.1 Energy use in residential buildings

A review on buildings' energy consumption has been presented by Perez-Lombard et al, 2008. The authors stated that HVAC systems contribute at 50 % of total building energy consumption. They also reported that in US dwellings the annual average energy consumption is at 147 KWh/m². Balaras et al, 2007 have reported on the energy consumption of the Hellenic building stock. The authors stated that the annual total energy consumption in European residential buildings varies between 150-230 KWh/m². However, in Eastern and Central European countries annual energy consumption in residential buildings is significantly higher at 250-400 KWh/m² than in Western EU countries. In Scandinavia well insulated buildings have annual energy consumption at 120-150 KWh/m² while the so-called low energy buildings at 60-80 KWh/m². According to the authors thermal and electrical energy consumption in residential buildings in Greece depends on the climatic zone and the year of their construction. Annual electrical energy consumption varies between 22.53 KWh/m² to 39.20 KWh/m² and annual thermal energy consumption between 52.1 KWh/m² to 189.9 KWh/m². Energy performance assessment of existing dwellings has been reported by Poel et al, 2007. The authors have presented a methodology and software to perform energy audits in buildings. They stated that in EU countries over 40 % of the final energy consumption is attributable to buildings while energy consumption in residential buildings corresponds at 63 % of total energy use in the building sector. Balaras et al, 2005 have reported on the heating energy consumption in European apartment buildings stating that the European annual average heat consumption is 174.3 KWh/m². The possibility of energy savings in Danish residential building stock has been reported by Tommerup et al, 2006. The authors indicated that energy saving measures are cost-effective and they could decrease the energy consumption and CO_2 emissions from residential buildings in this country. They concluded that there are no technical or economic barriers to hamper the improvement of energy performance in buildings. The only barrier, they mentioned, is the lack of knowledge and interest of the people to renovate their homes increasing their energy efficiency and sustainability.

2.2 Life cycle energy analysis in buildings

Energy is consumed in various stages over the lifespan of a building in the construction stage, during its operation, during its refurbishment and finally during its demolition. Ramesh et al, 2010 have presented an overview of life cycle energy analysis of various buildings. The authors analyzed data from 73 buildings across 13 countries concluding that operating energy corresponds at 80-90% of the life-cycle energy demand. Life cycle primary energy requirements in conventional residential buildings fall in the range of 150-400 KWh/m²yr. Suzuki et al, 1995 estimated the energy consumption and CO_2 emissions due to construction in different types of residential buildings in Japan. The authors found that depending on the type and construction of buildings their energy consumption varied between 833-2,777 KWh/m² and their CO_2 emissions between 250-850 $KgCO_2$/m². Hernandez et al, 2011 have presented a methodology for life-cycle building energy rating. The authors stated that current methods for assessing the energy performance of buildings account only for the operating energy use. However, the embodied energy should also be taken into account in order to have a more accurate energy classification of various buildings. Hernandez et al, 2010 have reported on defining life-cycle zero-energy buildings. According to them a life-cycle zero-energy building is a building whose primary energy use during its operation plus the energy embedded in materials and systems over its life span is equal to or less than the energy produced by renewable energy systems within the building.

2.3 Use of renewable energies in residential buildings

Various renewable energy technologies are currently used in residential buildings for heat and electricity generation including:

a) Solar thermal energy,
b) Solar photovoltaic energy,
c) Low enthalpy geothermal energy with heat pumps, and
d) Solid biomass

Applications of small wind turbines in residential buildings are currently rather limited. The necessary technologies for energy generation from renewable energy sources are reliable, mature and cost-effective. The high drop in prices of photovoltaic cells in the last few years has increased their penetration in the building sector. The recently introduced legal framework of net-metering allows the increased use of solar-PVs in buildings from environmentally conscious consumers. The technology of semi-transparent photovoltaic cells facilitates their integration in the building envelope. Although the majority of the currently used renewable energy technologies in buildings could produce on-site heat and cooling energy the most practical renewable energy technology which could generate electricity in private and public buildings is solar photovoltaic. Co-generation of heat and power using biomass as fuel is another option for carbon-free power generation. District heating using biomass or waste heat could also be considered for providing heat in buildings. Depending on availability of green energy sources, renewable energy technologies could replace fossil fuels in heat generation in buildings. Zero-carbon energy generation technologies which could be used in residential buildings are presented in Table 1.

Table 1. Sustainable energy technologies which could be used in residential buildings [1]

Renewable energy technology	Space heating	Space cooling	Electricity generation	Domestic hot water production
Solar thermal with flat plate collectors				+

Solar-PV			+	
Solid biomass burning	+			+
High efficiency Heat pumps	+	+		+
Small wind turbines			+	
Co-generation of heat and power using biomass	+		+	+
District heating using biomass	+			+
District heating using waste heat	+			+

[1] Source: Own estimations

2.4 Zero energy buildings

Li et al, 2013 have presented a review concerning zero-energy buildings (ZEBs). The authors reported that ZEBs involve two design strategies. The first concerns minimizing the need for energy use in buildings through energy-saving measures and the second the adoption of renewable energy technologies to meet the remaining energy needs. They concluded that ZEBs will play an important role in the future during the era of sustainable development. Torcellini et al, 2006 have presented a critical look at zero-energy buildings. The authors stated that a net zero-energy building is a residential or commercial building with significantly lower energy needs through energy efficiency gains such that its energy requirements can be covered with renewable energy technologies. The authors stated that there are four (4) commonly used definitions for zero-energy buildings:

a)   Net-zero site energy,
b)   Net-zero source energy,
c)   Net-zero costs, and
d)   Net-zero energy emissions.

A net-zero energy emissions building produces at least as much emissions-free renewable energy as it uses from emissions-producing energy sources. Marszal et al, 2011 have reviewed definitions and calculating methodologies of zero-energy buildings. The authors stated that the concept of a zero-energy building requires clear and consistent definition and a commonly agreed energy calculation methodology. Wang et al, 2009 have presented a case study in UK of a grid-connected zero-energy house. The house was using solar-PVs and wind turbines for electricity generation, a solar thermal system for hot water production and an under-floor heating system with a heat pump for space heating. Iqbal, 2004 has presented a feasibility study of a grid-connected zero-energy home in Newfoundland. The author stated that in St. John's Newfoundland the average annual wind speed is 6.7 m/s and he indicated that a wind turbine could provide all the required energy to the home. He also estimated that the cost of the energy system in order to convert the home to a zero-energy home is about 30 % of the total construction cost in a typical house in Newfoundland. Parker, 2009 has reported on perspectives of very low-energy homes in USA. He stated that very low-energy buildings can be easily achieved in North America. He suggested that energy efficiency measures should be a prerequisite to install a DHW system and a solar-PV system to near-zero energy homes. Musall et al, 2010 have presented an overview of net-zero energy solar buildings. They stated that the biggest challenge for all zero-energy building projects is the best fit of energy saving and renewable energy technologies used. For a small residential building they proposed the following combination of sustainable energy technologies:

a) A full passive house with solar thermal collectors,
b) Use of a heat pump,
c) Use of efficient electric appliances, and
d) A solar-PV system for electricity generation

Sartori et al, 2012 have reported on net zero energy buildings. The authors stated that although the concept of zero-energy buildings is generally understood, an internationally agreed definition is still lacking. They proposed the combined assessment of two criteria for zero-energy buildings. The first concerns the energy flaws exchanged between the building and the grid and the second the balance between on-site energy generation and energy consumption in the building. Zhu et al, 2009 have presented an energy and economic analysis on a zero-energy house versus a conventional house in USA. The authors found that various energy measures were cost-effective and among them high performing windows, energy saving lights, air-conditioners with a water-cooled condenser and a highly insulated roof. A solar-PV system and a solar thermal system installed in the zero-energy house had higher payback periods than the previous energy-saving technologies.

2.5 Zero CO$_2$ emission buildings due to energy use

Levine et al, 1996 have reported on mitigation options for CO$_2$ emissions in buildings. The authors stated that between 1971 and 1992 annual growth in CO$_2$ emissions from buildings varied widely ranging from 0.9 % in industrialized countries, 0.7 % in Eastern Europe and the former Soviet Union and 5.9 % in developing countries. A lot of CO$_2$ emissions in buildings are the result of population growth and growth in energy services particularly in developing countries. Policy instruments reducing energy use and CO$_2$ emissions in buildings include increase in energy prices, promotion of energy efficiency policies, technology transfer and increased research in innovative energy technologies used in buildings. Urge-Vorsatz et al, 2007 have presented an appraisal of policy instruments for reducing CO$_2$ emissions in buildings. The authors stated that currently the building sector contributes approximately at one third of energy-related CO$_2$ emissions worldwide while it is economically possible to achieve a 30 % emissions reduction. However, they claimed, numerous barriers prevent the realization of the high economic potential of energy and carbon reduction. Assessing many policy evaluation reports from various countries the authors concluded that the most cost-effective instruments achieving energy and CO$_2$ savings at negative costs for the society were the introduction of appropriate appliances standards, demand-side management programs and mandatory labeling. Urge-Vorsatz et al, 2008 have estimated the potential and cost of CO$_2$ mitigation in the world's buildings. Their findings showed that the highest potential for CO$_2$ mitigation in buildings in developing countries is associated with electricity savings in lighting and appliances use. The highest potential in developed countries and economies is related with energy savings. Vourdoubas, 2015 and Vourdoubas, 2016 presented the possibility of creating net zero-CO$_2$ emissions hotels and residential buildings due to energy use in Crete, Greece. The author proposed that two combinations of renewable energy technologies could cover all energy requirements in buildings zeroing their CO$_2$ emissions in Crete. The first combination included the use of solar thermal energy, solar-PV energy and solid biomass while the second the use of solar thermal energy, solar-PV energy and high efficiency heat pumps.

The purpose of current work is to present the realization of a small grid-connected residential building with zero-carbon footprint due to energy use which is located in Crete, Greece.

This was easily achieved only with the use of locally available renewable energies including solar energy and solid biomass. Their technologies for on-site heat and power generation are mature, reliable and cost-effective. Use of energy saving technologies in the building although desirable was not necessary. Economic and environmental analysis reveals that transformation of the grid-connected residential building in Crete, Greece to a zero-carbon emissions building with the use of the above-mentioned energy technologies is economically and environmentally attractive.

3. Differences between near zero energy buildings and near zero CO₂ emission buildings

EU directive 2010/31/EU promotes the creation of near-zero energy buildings (NZEBs) which are buildings with low energy consumption and low CO_2 emissions. Existing or new buildings could be transformed to NZEBs with the use of energy saving techniques and technologies which however, are in some cases costly. Their low energy load could be covered either with the use of conventional fuels or with the use of renewable energies. Depending on the energy source used they could emit zero, low or higher amounts of CO_2 in the atmosphere. On the contrary a zero-CO₂ emissions building (ZCO₂EB) does not necessarily reduce its energy consumption. However, it uses only renewable energies in order to cover its low or high energy loads and consequently it does not emit CO_2. A ZCO₂EB could utilize grid electricity derived from fossil fuels which though must be offset with green electricity over a year and the overall electricity balance must be zero. Therefore, the transformation of an existing building to ZCO₂EB does not necessarily require investments in energy-efficient (EE) technologies which are in some cases costly in order to reduce its overall energy consumption. However, it requires investments in renewable energy technologies (RET), preferably on-site, which are locally available, reliable, mature and cost effective. Currently EU finances the promotion of zero-CO₂ emissions buildings due to energy use through the INTERREG EUROPE program in order to improve current policies and to propagate good practices concerning the promotion of buildings with low or zero-energy consumption and CO_2 emissions. Although Greek legislation has complied with the European legislation regarding NZEBs there is currently no legal framework promoting ZCO$_2$EBs. The main differences between a NZEB and a ZCO$_2$EB are presented in Table 2.

Table 2. Differences between a NZEB and a NZCO$_2$EB [1]

	Nearly Zero Energy Building	Net Zero CO$_2$ Emissions Building
Energy consumption	Low	Not necessarily low
Energy sources used	Conventional fuels and/or renewable energies.	Only renewable energies. If fossil fuels are used they must be offset with renewable energies.
CO$_2$ emissions due to energy use	Low or zero.	Zero
Energy investments	In energy saving technologies and probably in renewable energies.	In renewable energies but not necessarily in energy saving technologies.
Cost effectiveness of the energy technologies used	Some energy saving technologies cost effective while others are not. Various R.E.T. are cost effective.	Various R.E.T. are cost effective.

[1] *Source: Own estimations*

4. Realization of a small residential building with zero CO$_2$ emissions due to energy use located in Crete, Greece

4.1 Building characteristics

The small residential building is located in the village of Gerani in the municipality of Platanias, in Western Crete, Greece (latitude 35.50.45 North and longitude 23.87.73 East). Its covered surface is 65 m^2, it is connected with the electric grid and a family with four members is currently living there. The building was constructed with reinforced concrete during 2001 and it was not properly thermally insulated. Greece is divided into four (4) climatic zones and the residential building is located in zone A with the mildest climate in the country. Average monthly air temperatures in this area vary from 10.8 °C in January to 26.6 °C in July. The annual solar irradiance is estimated at 1,738 KWh/m^2. The building envelope is not properly insulated but its doors and windows are energy-efficient constructed with double glazing.

4.2 Requirements for zeroing the carbon footprint in residential buildings

In order to zero the CO_2 emissions due to operational energy use in this building the following requirements must be fulfilled:

1. Fossil fuels must not be used for heat production in the building. They must be replaced with renewable energy sources, and
2. Grid electricity used must be offset annually with green electricity like solar-PV electricity. The electric grid of Crete is not interconnected with the country's continental grid and it is assumed that all grid electricity is generated with fossil fuels. However, currently in Crete, Greece renewable energies generate approximately 18 % of the annual generated electricity in the island.

4.3 Energy loads

Energy is used in the building in various sectors including space heating, space cooling, hot water production, lighting and operation of various electric appliances. Overall annual energy consumption in the residential building is estimated at 180 KWh/m^2 (11,700 KWh). The estimated energy consumption per sector is presented in Table 3.

Table 3. Annual energy consumption in various sectors in the residential building [1]

Sector	% of total energy used	Specific energy consumption (KWh/m^2)	Annual energy consumption (KWh)
Space heating	63	113.4	7,371
Hot water production	9	16.2	1,053
Heating energy	72	129.6	8,424
Lighting	12	21.6	1,404
Operation of	16	28.8	1,872

various appliances including space cooling			
Electricity	28	50.4	3,276
Total operating energy	100	180	11,700
Embodied energy [2]	17.65	31.76	2,064.7
Total energy consumption including the embodied energy [2]	117.65	211.76	13,764.7

[1]*Covered area 65 m², ²assuming that operating energy corresponds at 85 % and embodied energy at 15 % of the total life cycle energy consumption*

According to table 3 the total amount of energy used for heating in the residential building is significantly higher than the amount used for electricity.

4.4 Use of renewable energy technologies

Three different renewable energy technologies were used in the residential building. They included a solar thermal system for domestic hot water production, a solar photovoltaic system for electricity generation and a solid biomass burning system for space heating. The solar thermal system was installed on the roof terrace and the area of its solar collectors was 2 m² corresponding at 1.4 KW$_{th}$. It is expected to cover approximately 85 % of the annual needs in DHW in the residential building. The solar photovoltaic system had a nominal power at 3 KW$_p$ and its annual electricity generation has been estimated at 4,500 KWh. It was installed on the roof terrace of the grid connected building according to net-metering regulations. Since the annual electricity consumption in the building was expected to be lower than 4,500 KWh the excess generated electricity from the solar-PV system would be injected into the grid. A wood stove at 6 KW$_{th}$ was installed in the building utilizing olive tree wood for space heating and providing hot water when needed. The annual wood consumption was estimated at 3 tons and its energy efficiency was not exceeding 60 %. Solid biomass is expected to cover all the space heating needs of the building during the winter. All renewable energy systems used were located on-site.

4.5 Energy balance in buildings

Heat and electricity generation from the renewable energy systems installed in the building has been estimated as follows:

a) Heating value of the solid biomass used: 3,600 Kcal/kg, with thermal efficiency of the wood stove 60 % and annual operation 600 hours,
b) Annual heat generation from the solar thermal system: 860 KWh/KW$_{th}$, and
c) Electricity generation from the solar-PV: 1,500 KWh/KW$_p$.

Energy generation from the renewable energy systems in the residential building as well as the energy balance is presented in Table 4.

Table 4. Energy generation and energy balance in the residential building

Energy system	Power of the energy system	Annual energy generation (KWh/yr)	Annual energy consumption (KWh/yr)	Energy surplus (KWh/yr)
Solar thermal	1.4 KW$_{th}$	1,204	1,053	151
Biomass burning	6 KW$_{th}$	9,031	7,371	1,660
Heating energy		10,235	8,424	1,811
Solar-PV (Electricity)	3 KW$_{el}$	4,500	3,276	1,224
Total	7.4 KW$_{th}$ +3 KW$_{el}$	7.4 KW$_{th}$ + 3 KW$_{el}$	11,700	3,035

Energy generation from the above-mentioned renewable energy systems according to Table 4 is higher than the average energy consumption in the building in order to cope with unexpected variations or failures. The over-design of the energy systems results in slightly higher investment costs and it should be noted that according to current regulations of net-metering in Greece the surplus electricity generated is injected into the grid without any financial compensation. The annual energy surplus in the building at 3,035 KWh/yr exceeds its embodied energy which is 2,064.7 KWh/yr.

5. Cost analysis

The capital cost of the three renewable energy systems installed in the residential building has been estimated as well as the annual solid biomass cost. The capital cost of the solar thermal system at 1.4 KW$_{th}$ is 1,000 € the unit cost of the biomass burning system is 500 €/KW$_{th}$ and the unit cost of the solar-PV system is

1,700 €/KW$_p$. Current cost of solid biomass in Crete is 0.13 €/kg. Cost estimations are presented in Table 5.

Table 5. Cost estimations of the renewable energy systems installed in the small residential building in Crete

Energy system	Power	Capital cost (€)	Operating cost (€/yr)[1]
Solar thermal	1.4 KW$_{th}$	1,000	0
Solar-PV	3 KW$_{el}$	5,100	0
Solid biomass burning	6 KW$_{th}$	3,000	390
Total		9,100	390

[1] Maintenance and depreciation costs have not been included

Assuming that the current construction cost of the small residential building in Crete is approximately 84,500 € (1,300 €/m²) the total installation cost of the above-mentioned renewable energy systems corresponds at 10.77 % of its overall construction cost. Since heating oil has been replaced by woody biomass while the electricity as well as the hot water used in the building are generated on-site by solar energy significant annual cost savings concerning the energy bill have been obtained.

6. Environmental analysis

Since renewable energies have replaced totally the use of conventional fuels in the residential building its CO_2 emissions due to energy use have been zeroed. In order to estimate the CO_2 emissions saved it is assumed that initially the building was covering all its energy needs with electricity and heating oil. Electricity was used for space cooling, hot water production and operation of various electric appliances while heating oil was used for space heating. CO_2 emissions due to conventional fuels use are presented in Table 6.

Table 6. CO_2 emissions due to conventional fuels use in a small residential building in Crete[1]

Fuel	Use	Quantity (KWh/yr)	CO_2 emissions (KgCO_2/year)	CO_2 emissions (KgCO_2/m²year)
Electricity consumption	Cooling, hot water production and operation	4,329	3,247	49.95

	of electric devices			
Heating oil consumption	Space heating	7,371	2,257	34.72
Total energy		11,700	5,504	84.67

[1]*Electricity emission coefficient, 0.75 kg CO_2/KWh, Heating oil emission coefficient, 3.2 kg CO_2/kg of oil, Net heating value of diesel oil, 9,000 kcal/kg*

According to Table 6, CO_2 emissions in the building due to electricity use are higher than emissions due to heating oil use. Therefore, the annual total CO_2 emissions savings due to the use of renewable energies in the building are 5,504 KgCO_2. The required capital investments in the above-mentioned renewable energy systems in order to achieve the goal of a net zero carbon emissions residential building due to energy use are 1.65 € per kgCO_2 saved annually.

7. Discussion

Reducing or zeroing CO_2 emissions due to energy use in residential buildings is of paramount importance for mitigating climate change. Zero-CO_2 emissions buildings could be achieved by increasing their energy efficiency and replacing conventional fuels use with renewable energies. The methodology used for realization of the small residential building with zero CO_2 emissions due to energy use is based in the followings. a) Definition of the criteria for achieving a zero CO_2 emission building, b) Estimating the energy demand in the building, and c) Estimating the locally available and cost effective renewable energies which could cover the energy demand in the building. The abovementioned methodology assumes that a) All grid electricity is generated with fossil fuels, and b) The solar-PV system will operate smoothly without failures providing continuously electricity into the grid. It is obvious that other combinations of renewable energy technologies apart from those mentioned in the current study could achieve the same result in the residential building. Mussal et al, 2010 have presented another combination of renewable energy technologies which also results to a net-zero energy solar building. The building was characterized by a passive solar system with solar thermal collectors, a heat pump, efficient electric appliances and a solar-PV system. Another combination of different renewable energy technologies for achieving the same result has been proposed by Wang et al, 2009. The authors proposed, for a grid-connected zero-energy house in UK, solar-PVs and a wind turbine for electricity generation, a solar thermal system for hot water production and a heat pump for space heating. The estimated cost of the required renewable energy technologies in order to transform the residential building into a NZCO$_2$EB corresponds at 10.77 % of its overall construction cost which is significantly lower than the required cost reported by Iqbal, 2004 for a grid connected zero energy home in Newfoundland which has been estimated at 30 % of its construction cost. In each territory the availability of various renewable energies combined with their cost effectiveness and the size of the buildings would indicate their optimum combination in order to zero their net carbon footprint. It should be pointed out that for a small building in Mediterranean region including the island of Crete, Greece the best renewable energy technology for on-site electricity generation is the solar-PV technology. Wind turbines are suitable and cost effective only in the case that the annual wind speeds on-site are high. On the contrary there are more available renewable energy technologies for heat generation. In the current building a different combination of renewable energy technologies for zeroing its carbon emissions could be: a) A solar thermal system for hot water production, b) A high efficiency heat pump for space heating and cooling, and c) A solar-PV system for electricity generation. The possibility of zeroing the carbon footprint of the building over its life cycle should be also examined. In this case its embodied energy should be offset with green electricity. The excess solar electricity injected into the grid over its life time should be equal to its embodied energy. However, the existing legal framework for net-metering in Greece does not foresee financial compensation in the case that more electricity is injected annually into the grid than the building's electricity consumption. This fact discourages the building owners to install solar-PV systems with higher nominal capacities in order to create positive energy buildings and to receive financial compensation for that.

8. Conclusions

Realization of a small, grid-connected, residential building with net zero-CO$_2$ emissions due to energy use in Crete, Greece has been implemented rather easily. The technical and economic feasibility has been proved as well as the environmental advantages. This indicates that there are not technical, economic or legal barriers for achieving net zero-CO$_2$ emission buildings due to operating energy use. However, incentives should be provided to public and private sector in order to promote this type of buildings combined with sensitization of citizens. The creation of various demonstrative municipal buildings with net zero CO$_2$ emissions using different renewable energy technologies could be a good example in order to familiarize the local society with them. On-site electricity generation requires the use of a solar-PV system while the net-metering initiative which was legalized in Greece in the last two years allows that. The high solar irradiance in the island of Crete and the sharp drop in the prices of solar-PV cells make their use very attractive. Alternative renewable energy technologies for power generation in small residential buildings include the use of small wind turbines and biomass co-generation systems which are not appropriate in the current case. The energy efficiency in the abovementioned building has not been improved with proper thermal insulation but all its energy needs have been covered with the use of solar energy and solid biomass. The covered area of the building was 65 m^2, its annual specific energy consumption 180 KWh/m^2 and its total annual operating energy consumption 11,700 KWh/m^2. Renewable energies which are abundant in Crete were used. The applied technologies included a solar thermal system for hot water production, a solar-PV system for electricity generation and a solid biomass burning system for space heating. Heat and electricity generated on-site were slightly higher than the energy consumption in the building and electricity injected into the grid was also higher than the annually consumed electricity. The capital cost of the above-mentioned renewable energy systems was estimated at 9,100 € which corresponds at 10.77 % of the total construction cost of the building. The achieved annual CO_2 emissions savings have been estimated at 5,504 kgCO_2 or 84.67 kgCO_2/m^2. According to various studies the building's embodied energy corresponds approximately at 15 % of its total operating energy during its life cycle. The embodied energy of the building could be offset by increasing slightly the size of the solar-PV system and injecting annually surplus electricity into the grid. The proposed technologies for zeroing CO_2 emissions are reliable, mature and cost-effective. Improvement of its energy performance with better thermal insulation of the walls, doors, windows and the roof could result in smaller sizing of required renewable energy systems. Further work regarding the creation of NZCO$_2$EBs in Crete is necessary considering first the reduction of their energy consumption and afterwards the use of renewable energy technologies. The future realization of a NZCO$_2$EB in Crete, Greece using only solar energy and geothermal heat pumps for covering all its annual energy needs is also desirable in order to compare its performance with the building already studied.

References

[1] Balaras, C.A., Droutsa, K., Dascalaki, E. & Kontoyiannidis, S. (2005). "Heating energy consumption and resulting environmental impact of European apartments buildings", *Energy and Buildings*, 37, pp. 429-442. doi:10.1016/j.enbuild.2004.08.003.

[2] Balaras, C.A., Gaglia, A.G., Georgopoulou, E., Mirasgedis, S., Sarafidis, G. & Lalas, D.P. (2007). "European residential buildings and empirical assessment of the Hellenic building stock, energy consumption, emissions and potential energy savings", *Building and Environment*, 42, pp. 1298-1314. doi:10.1016/j.buildenv.2005.11.001.

[3] EU directive 2010/31/EU, on the energy performance of buildings, 19/5/2010, retrieved on 21/6/2017 from http://eur-lex.europa.eu/LexUriServ/LexUriServ.do?uri=OJ:L:2010:153:0013:0035:en:PDF

[4] Hernandez, P. & Kenny, P. (2010). "From net zero to zero energy buildings. Defining life cycle zero energy buildings (LC-ZEB)", *Energy and Buildings*, 42, pp. 815-821. https://doi.org/10.1016/j.enbuild.2009.12.001.

[5] Hernandez, P. & Kenny, P. (2011). "Development of a methodology for life cycle building energy ratings", *Energy Policy*, 39, pp. 3779-3788. *doi*:10.1016/j.enpol.2011.04.006.

[6] Interreg Europe project "Promotion of zero CO_2 emission buildings due to energy use". Retrieved on 21/6/2017 from https://www.interregeurope.eu/zeroco2/

[7] Iqbal, M.T. (2004). "A feasibility study of a zero energy home in Newfoundland", *Renewable Energy*, 29, pp. 277-289. doi: 10.1016/S0960-1481(03)00192-7.

[8] Levine, M.D., Price, L. & Martin, N. (1996). "Mitigation options for carbon dioxide emissions from buildings", *Energy Policy*, 24(10-11), pp. 937-949. doi:10.1016/s0301-4215(96)80359-4.

[9] Li, D.H.W., Yang, L. & Lam, J.C. (2013). "Zero energy buildings and sustainable development implications - A review", *Energy*, 54, pp. 1-10. doi: 10.1016/j.energy.2013.01.070.

[10] Marszal, A.J., Heiselberg, P., Bourrelle, J.S., Mussal, E., Voss, K., Sartori, J. & Napolitano, A. (2011). "Zero energy building - A review of definitions and calculation methodologies", *Energy and Buildings*, 43, pp. 971-979. https://doi.org/10.1016/j.enbuild.2010.12.022.

[11] Mussal, E., Weiss, T., Voss, K., Lenoir, A., Donn, M., Cory, S. & Garde, F. (2010). "Net zero solar buildings; An overview and analysis on worldwide building

projects", *in conference proceedings in EuroSun 2010, International solar energy society*, Graz, Austria, 20/9-1/10. doi:10.18086/eurosun.2010.06.16

[12] Parker, D.S. (2009). "Very low energy homes in the United States: Perspectives on performance from measured data", *Energy and Buildings*, 41, pp. 512-520. http://dx.*doi*.org/10.1016/ j.enbuild.2008.11.017.

[13] Perez-Lombard, L., Ortiz, J. & Pout, Chr. (2008). "A review on buildings energy consumption information", *Energy and Buildings*, 40, pp. 394-398. *doi*:10.1016/*j*.enbuild.2007.03.007.

[14] Poel, B., Cruchten, G.V. & Balaras, A. (2007). "Energy performance assessment of existing dwellings", *Energy and Buildings*, 39, pp. 393-403. doi:10.1016/j.enbuild.2006.08.008.

[15] Ramesh, T., Prakash, R. & Shukla, K.K. (2010). "Life cycle energy analysis of buildings: An overview", *Energy and Buildings*, 42, pp. 1592-1600. doi:10.1016/j.enbuild.2010.05.007.

[16] Sartori, I., Napolitano, A. & Voss, K. (2012). "Net zero energy buildings: A consistent definition framework", *Energy and Buildings*, 48, pp. 220-232. doi:10.1016/j.enbuild.2012.01.0.

[17] Suzuki, M., Oka, T. & Okada, K. (1995). "The estimation of energy consumption and CO_2 emissions due to housing construction in Japan", *Energy and Buildings*, 22, pp. 165-169. doi:10.1016/0378-7788(95)00914-j.

[18] Tommerup, H. & Svendsen, S. (2006). "Energy savings in Danish residential buildings stock", *Energy and Buildings*, 38(6), pp. 618-626. DOI: 10.1016/j.enbuild.2005.08.017.

[19] Torcellini, P., Pless, S., Deru, M. & Crawley, D. (2006). "Zero energy buildings: A critical look at the definition", *Presented at ACEEE summer study*, Pacific Grove, California, USA, 14-18/8/2006.

[20] Urge-Vorsatz, D. & Novikova, A. (2008). "Potentials and costs of carbon dioxide mitigation in the worlds buildings", *Energy Policy*, 36, pp. 642-661. doi:10.1016/j.enpol.2007.10.009.

[21] Urge-Vorsatz, D., Koeppel, S. & Mirasgedis, S. (2007). "Appraisal of policy instruments for reducing building CO_2 emissions", *Building Research and Information*, 35(4), pp. 458-477. http://dx.doi.org/10.1080/09613210701327384.

[22] Vourdoubas, J. (2015). "Creation of hotels with zero CO_2 emissions due to energy use: A case study in Crete-Greece", *Journal of Energy and Power Sources*, 2(8), pp. 301-307.

[23] Vourdoubas, J. (2016). "Creation of zero CO_2 emission buildings due to energy use. A case study in Crete-Greece", *Journal of Civil Engineering and Architecture Research,* 3(2), pp. 1251-1259.

[24] Wang, L., Gwilliam, J. & Jones, P. (2009). "Case study of zero energy house in U.K.", *Energy and Buildings,* 41, pp. 1215-1222. doi:10.1016/j.enbuild.2009.07.001.

[25] Zhu, L., Hurt, R., Correa, D. & Boehm, R. (2009). "Comprehensive energy and economic analysis on a zero energy house versus a conventional house", *Energy,* 34, pp. 1043-1053. https://doi.org/10.1016/j.energy.2009.03.010.

[3] Creation of zero CO$_2$ emissions residential buildings due to operating and embodied energy use in the island of Crete, Greece

1. Introduction

Energy consumption in buildings corresponds at 40 % of the total energy consumption in EU while it is responsible for undesired GHG emissions. Improving building's energy efficiency is of paramount importance for promoting energy and environmental sustainability while current European regulations are targeting at nearly zero energy buildings focusing on their operating energy. Apart from their operating energy buildings utilize energy during their construction, maintenance, refurbishment and demolition which is called embodied energy. However, the concept of zero CO$_2$ emissions buildings due to operating and embodied energy use has not been developed so far, neither there are official legal regulations promoting this type of buildings. Current research investigates the possibility of using mature, reliable and cost-effective renewable energy technologies for covering all the operating energy needs in residential buildings taking also into account their embodied energy. Replacement of fossil fuels use with renewable energies in buildings could result in lowering or zeroing their carbon emissions contributing in global efforts for decreasing GHG emissions. Present research is important because it indicates the way that the use of various renewable energy technologies in residential buildings could zero their CO$_2$ emissions due to both operating and embodied energy use.

2. Literature survey

2.1 Low energy buildings

European legislation is forcing towards creation of new buildings with nearly zero energy consumption and refurbishment of old buildings in order to reduce their energy use (Directive 2002/91/EC, Directive 2010/31/EU). Torcellini et al, 2006 [1] have reported on various definitions of zero energy buildings. The authors reported four definitions as follows: net zero site energy, net zero source energy, net zero energy costs and net zero energy emissions. A net zero energy building, according to them, produces at least as much emissions-free renewable energy as it uses from emissions-producing energy sources. Marszal et al, 2011 [2] have presented a review of definitions and calculation methodologies for zero energy buildings. The authors stated that the zero energy building concept requires a clear and consistent definition and a common agreed energy calculation methodology. They discussed various issues and approaches which should be clarified in order to facilitate a consistent and common agreed definition of zero energy building. Current published literature, according to author's knowledge, has not included so far the embodied energy in the calculation of zero energy buildings. Vourdoubas, 2016 [3] has reported on the creation of zero CO_2 emissions residential buildings due to operating energy use in Crete, Greece. The author estimated, for a typical residential building in Crete, Greece, its annual CO_2 emissions due to operating energy use at 84.55 $kgCO_2/m^2$. Realization of a small residential building with zero CO_2 emissions due to energy use in Crete, Greece has been reported from Vourdoubas, 2017 [4]. The author stated that renewable energy sources including solar thermal energy, solar-PV energy and solid biomass were used for covering all its energy needs. The cost of the required renewable energy systems corresponded at 10.77 % of its overall construction cost. He also estimated that it was equal at 1.65 € per $kgCO_2$ saved annually in the building. Tselepis, 2015 [5] has implemented various case studies, including studies in residential buildings, concerning electricity generation with solar-PVs in Greece according to net-metering regulations, indicating that it was profitable.

2.2 Environmental impacts of buildings due to energy use

A review of current trends in operating versus embodied energy in buildings has been published by Ibn-Mohammed et al, 2013 [6]. The authors stated that in order to mitigate climate change, buildings must be designed and constructed with minimum environmental impacts. The total life cycle emissions from buildings are due to operating and embodied energy use. Considerable efforts have been made to reduce operating emissions from buildings but little attention has been paid to embodied emissions. Therefore, a critical review on the relation between operating and embodied emissions is necessary in order to highlight the importance of embodied emissions. Estimation of energy consumption and CO_2 emissions due to housing construction in Japan has been calculated by Suzuki et al, 1995 [7]. Energy consumption at 3-10 GJ/m^2 (833-2,777 KWh/m^2) has been found depending on the type and construction of the building while its CO_2 emissions during the construction stage varied at 250-850 kg/m^2. Guellar-Franca et al, 2012 [8] have investigated the environmental impacts in UK residential sector. The authors reported that over a period of 50 years, 90 % of the global warming potential of the residential buildings is due to their operating carbon emissions and only 10 % is due to their embodied carbon emissions. Syngros et al, 2016 [9] have reported on embodied CO_2 emissions in building construction materials in Hellenic dwellings. The authors analyzed CO_2 emissions corresponded to construction materials for four typical dwellings in Greece, estimating their average CO_2 emissions at 777 $KgCO_2/m^2$. Over a life span of 50 years the annual embodied CO_2 emissions corresponded to construction materials were at 15.54 $kgCO_2/m^2$. The authors also stated that emissions were mainly due to construction materials while the share of the electro-mechanical installations was below 2 %.

2.3 Energy consumption in buildings

Karimpour et al, 2014 [10] have reported on minimizing the life cycle energy in buildings. The authors stated that in mild climates embodied energy in buildings represents up to 25 % of the total life cycle energy. In the future, when operating energy in buildings will be reduced due to construction of nearly zero energy buildings, the ratio of embodied energy to total life cycle energy will be increased. The need for an embodied energy measurement protocol for buildings has been reported by Dixit et al, 2012 [11]. The authors stated that studies have revealed the growing significance of embodied energy in buildings. However, current estimations of embodied energy are unclear and vary greatly. The authors recommended an approach to derive guidelines that can be developed into a globally accepted protocol. Cellura et al, 2014 [12] reported on the operating and embodied energy of an Italian building. The authors emphasized the key issue of embodied energy in the building which is particularly important in the case of low energy buildings. They also pinpointed the difficulty in defining the reference area in the building, including service and unheated zones as well as the absence of an internationally accepted protocol for that. Dixit at el, 2010 [13] have presented a literature review regarding the identification of parameters for embodied energy measurements in buildings. The authors stated that current methods of estimation are inaccurate and unclear. They concluded in a set of parameters that differ and cause variation and inconsistency in embodied energy estimation. Berggren et al, 2013 [14] have reported on life cycle energy analysis in buildings. With reference to net zero energy buildings, where on-site renewable energy generation covers the annual energy load, the authors analyzed the increase of embodied energy compared with the decrease in operating energy. They concluded that: a) in the last decades the embodied energy in new buildings has slightly decreased, b) the relationship between embodied energy and life cycle energy use is almost linear, and c) the relative share of embodied energy to life cycle energy use has increased. A methodology for life cycle building energy rating has been developed by Hernandez et al, 2011 [15]. Apart from the operating energy used during its operation, the embodied energy consumed during the construction of the building usually is not taken into account. For buildings with "net zero energy use" during their operation the only life-cycle energy is the embodied energy. In a case study of a detached house in Ireland the authors have estimated its embodied energy at 1,000 KWh/m^2. A review of the life cycle energy in conventional and low energy buildings has been reported by Sartori et al, 2007 [16]. The authors have analyzed 60 cases found in literature which showed that operating energy represents the largest part of total energy demand. They stated that there is a linear relation between operating and total energy use, concluding that low energy buildings are more energy-efficient although their embodied energy is slightly higher. Ramesh et al, 2010 [17] have presented an overview on life cycle energy analysis in buildings. The authors analyzed the results of 73 cases from 13 countries including residential and office buildings. They found that operating energy corresponds at 80-90 % of the life cycle energy use and the rest corresponds to embodied energy. They estimated the life cycle primary energy requirements at 150-400 KWh/m^{2}year for residential buildings and at 250-550 KWh/m^{2}year for offices. The authors also mentioned that low energy buildings perform better than zero operating energy buildings in the life cycle context. Adalberth et al, 2001 [18] have reported on life cycle energy assessment, regarding environmental impacts, of four multi-family residential buildings located in Sweden built in 1996. The authors stated that environmental impacts during operation of the buildings contributed at 70-90 % of their total life cycle impacts. Since environmental impacts are directly related with their energy consumption, they suggested that design and construction of energy-efficient buildings, with low energy related carbon emissions, results in minimizing their environmental impacts. A report on life cycle energy and environmental performance of a new university building has been made by Scheuer et al, 2003 [19]. The authors conducted a life cycle assessment of a 7,300 m^2 six-story building located in the University of Michigan campus. The primary annual energy intensity over a life time of 75 years was estimated at 1,171 KWh/m^2 and the operating energy accounted at 94.4 % of its life cycle energy use. Ramesh et al, 2013 [20] have reported on a case study of life cycle energy analysis of a multifamily residential house in India. The authors estimated that operating energy in the house corresponded at 89 % of its life cycle energy use and the remaining 11 % corresponded to embodied energy. The primary annual energy consumption was estimated at 288 KWh/m^2 and the authors mentioned that building-integrated solar-PVs is an attractive option for the reduction of life cycle energy use. Fay et al, 2000 [21] have reported on life cycle energy analysis of buildings in Australia. The authors stated that as operating energy in buildings is decreasing due to energy efficiency improvements, embodied energy becomes more significant. They also mentioned that while a net zero operating energy building is now achievable, a net zero life cycle energy building is likely to be more difficult.

The aim of the current study is to investigate the possibility of creating zero CO_2 emissions grid-connected residential buildings due to operating and embodied energy use with reference to the island of Crete, Greece. Appraisal of operating and embodied energy needs in the building was conducted first, followed by sizing the solar-PV systems in order to zero the overall CO_2 emissions due to both operating and embodied energy use. The current study indicates the way that renewable energy technologies can be used in residential buildings in order to zero their overall carbon emissions due to operating and embodied energy use with reference to the island of Crete, Greece.

3. Operating and embodied energy in buildings

Energy is consumed in various phases during the life cycle of a building including:
a) Energy consumed during its construction,
b) Energy consumed during its operation,
c) Energy consumed during its refurbishment, and
d) Energy consumed during its demolition.

The sum of energies consumed during its construction, refurbishments and demolition is considered as the embodied energy in the building. Although there is a lack of a generally accepted methodology for the estimation of the embodied energy in various types of buildings results of many studies implemented worldwide have shown that for conventional buildings the embodied energy corresponds approximately at 15 % of the total life cycle energy use. Therefore, operating energy has the largest share in total life cycle energy use. The necessity to cope with climate change and the efforts to reduce GHG emissions have altered the way that new buildings are constructed and old buildings are energy-renovated. Use of various cost-effective renewable energy technologies in buildings is also promoted resulting in reduced carbon emissions due to energy use. Apart from Europe in many other countries creation of energy-efficient buildings is of high priority. Creation of buildings with high energy efficiency requires use of new materials which have more embodied energy. Therefore, new buildings which have nearly zero energy consumption have also lower share of operating energy in their total life cycle energy use. In new buildings the embodied energy will have an increased share in their total energy use which could exceed 25 % compared with the current 15 %. Therefore, the significance of embodied energy in total life cycle energy use in new and in energy-renovated buildings is expected to increase in the future.

4. Zero CO_2 emission residential buildings due to operating energy use

Residential buildings consume energy for space heating and cooling, hot water production, lighting and operation of various electric appliances. Total and per sector energy consumption in buildings depends on many parameters including the type of building construction, local climate, occupants' behavior etc. A high quality constructed building having proper thermal insulation requires less operating energy and emits less CO_2 due to energy use. Therefore, its transformation to zero CO_2 emission building is easier compared with a lower quality constructed building. Vourdoubas, 2016 [3] has reported that a grid-connected residential building could zero its CO_2 emissions due to energy use if the following two conditions are fulfilled:

a) Fossil fuels used for space heating and hot water production would be replaced by renewable energy sources, and

b) Grid electricity used annually in the operation of the building would be offset by solar-PV electricity, generated on-site with photovoltaic panels.

Energy used for space heating and solar water production in a residential building can be produced with solid biomass, solar thermal energy and heat pumps. Solar energy, solid biomass and low enthalpy geothermal energy are available in many territories in Southern Europe while the energy generation technologies are reliable, mature and cost-effective. In many countries annual grid electricity used in a building can be offset with solar-PV generated electricity and injected into the grid according to a net-metering initiative. In southern European countries solar-PVs are not only used in buildings but also for electricity generation, in stand-alone systems, injected directly into the grid. The current drop in prices of solar-PV panels has increased their cost effectiveness and their attractiveness regarding their installation on the roofs of grid-connected residential buildings, generating annually part or all of their electricity needs, obtaining at the same time financial compensation. However, if the installed solar-PV system will annually generate and inject into the grid more electricity than the actual annual consumption in the building, the excess electricity is not financially compensated. In fact the current legal framework for net-metering in many countries does not exclude the installation of solar-PVs in residential buildings, generating annually more electricity than the actual consumption in the building. In this case though there is no financial compensation for the excess electricity injected into the grid. The renewable energy sources and the required technologies for creation of zero CO_2 emissions residential buildings are presented in Table 1.

Table 1. Renewable energy sources and their technologies which could be used in Crete in order to zero CO_2 emissions due to energy use in buildings [1]

	Renewable energy source	Technology used	Generated energy
1.	Solar energy	Solar thermal	Hot water
2.	Solar energy	Solar photovoltaic	Electricity
3.	Solid biomass	Biomass burning	Heat for space heating and hot water production
4.	Low enthalpy geothermal energy	Heat pumps	Heat and cooling

[1] Source: Own estimations

5. Zero CO_2 emission residential buildings due to operating and embodied energy use

In order to zero CO_2 emissions due to life cycle energy use in a residential building, an additional requirement to those in section 4 must be fulfilled as follows:

c) An additional solar-PV system installed on site must generate annually and inject into the grid an amount of electricity equal to its total embodied energy, divided by the life span of the building.

With reference to a residential building located in Crete, Greece the nominal power of the required solar-PV system is estimated in two cases. In the first case, additionally to solar energy, solid biomass is used for space heating, while in the second, a high efficiency heat pump. The sizing of the solar-PV system is made for offsetting its operating electricity and its embodied energy as well. In the following analysis three dimensionless parameters have been used including: a) the percentage of embodied energy to life cycle energy use in the building, b) the coefficient of performance (C.O.P.) of a heat pump, and c) the share of grid electricity generated by non-CO_2 emitting fuels. The first parameter varies depending on the construction of the building the behavior of the occupants and the local climate while an indicative value has been taken from existing literature data. The value of the second parameter is usual in commercial heat pumps and the third parameter depends on the electric grid and the energy mix used. The value taken in the following estimations is currently representative for Crete, Greece.

5.1 Mathematical formulation

The following equations were used:
X1= energy consumed for space heating in the building (KWh/m^{2}year)
X2= energy consumed for hot water production (KWh/m^{2}year)
X3= energy consumed for lighting (KWh/m^{2}year)
X4= energy consumed for the operation of various electric appliances (KWh/m^{2}year)
Xoper = operating energy in the building (KWh/m^{2}year)
Xemb= embodied energy in the building (KWh/m^{2}year)
X li.cy. = life cycle energy use (KWh/m^{2}year)
a= percentage of embodied energy to life cycle energy use (dimensionless number)
z= Annual energy generation from a solar-PV system (KWh/KW$_p$)
C.O.P. = coefficient of performance of a heat pump (dimensionless number)
P= Nominal power of a solar-PV system (KW$_p$)

Xoper = X1+X2+X3+X4   [1]
Xli.cy. = X1+X2+X3+X4+Xemb   [2]
Xli.cy. =Xoper+Xemb   [3]
Xemb. =a*Xli.cy.=a*[Xoper+Xemb] = a*Xoper+a*Xemb,   [4]
Xemb =Xoper * a/(1-a)   [5]

In the case that electricity is not consumed for space heating and for hot water production in the building, then
P= (X3+X4+Xemb)/z (KW$_p$/m^2)   [6]

In the case that a heat pump is used for space heating and electricity is not consumed for hot water production in the building, then

$$P = (X1/C.O.P. + X3 + X4 + Xemb)/z \quad (KW_p/m^2) \quad [7]$$

5.2 Adjustment due to grid electricity generation from renewable energies and nuclear power

Part of grid electricity is generated by non-CO_2 emitting fuels like renewable energy sources and nuclear power. In this case the solar-PV generated electricity should offset only the grid electricity generated by fossil fuels.
If Y % is the share of grid electricity generated by non-CO_2 emitting fuels, then equations 6 and 7 can be written as follows:

$$P1 = ((100-Y)/100) * (X3 + X4 + Xemb)/z \quad [6A]$$

and

$$P2 = ((100-Y)/100) * ((X1/C.O.P.) + X3 + X4 + Xemb)/z \quad [6B]$$

where P1 and P2 correspond to the nominal power of the required solar-PV systems adjusted for the share of non-CO_2 emitting fuels used in grid electricity generation.

5.3 Estimation of the required solar-PV system when solid biomass is used for space heating

Energy consumption per sector for a residential building located in Crete, Greece is presented in Table 2.

Table 2. Typical distribution of operating energy use in a residential building in Crete, Greece [1]

Sector	% of energy used	Annual energy consumption (KWh/m^2)
Space heating	63	107.1
Hot water production	9	15.3
Lighting	12	20.4
Operation of various appliances including space cooling	16	27.2
Total	100	170

[1] *Vourdoubas, 2016*

It should be mentioned that, although the climate in Crete is mild, the high share of energy use for space heating in the residential building is due to the fact that most buildings constructed before 2010 in Greece have poor thermal insulation. In the following estimations it has been assumed that the embodied energy in the residential building with a life span of 50 years is equal at 15 % of its life cycle energy use (Ramesh et al, 2010 [17]). This is an average value reported also in other studies concerning life cycle consumption in buildings.

Setting in equation 5, Xoper = 170 KWh/m^{2}year, and a=0.15, then Xemb = 30 KWh/m^{2}year

Setting in equation 6, X3 = 20.4 KWh/m^{2}year, X4 = 27.2 KWh/m^{2}year, then Xemb = 30 KWh/m^{2}year

while taken into account that the annual electricity generation from solar-PVs in Crete, Greece is approximately 1,500 KWh/KW$_p$,

z = 1,500 KWh/year per KW$_p$, Y=0.18

It is estimated that currently in Crete, Greece, 18 % of grid electricity is generated from renewable energy sources mainly from wind and solar-PV energy,

Therefore, P = 0.052 KWp/m^2 and P1= 0.042 KWp/m^2

Therefore, the nominal power of the required solar-PV system which will generate annually and inject into the grid the amount of electricity used for lighting, operation of electric appliances as well as for annual compensation of the embodied energy of the building over a period of 50 years, is 0.042 KWp/m^2. For a residential building with a covered surface of 100 m^2, it is 4.2 KW$_p$.

Assuming that the solar-PV system will generate annually only the electricity used for the operation of the building (Xemb= 0), equation 6A gives P1=0.026 KWp/m^2. For a residential building with a covered area at 100 m^2, the nominal power of the required solar-PV system is 2.6 KWp. The estimated additional nominal power of the solar-PV system is 1.6 KWp. The additional solar-PV system will generate and inject into the grid, over a period of 50 years, electricity equal to the embodied energy in the building.

5.4 Estimation of the required solar-PV system when a high efficiency heat pump is used for space heating

When a heat pump is used for space heating in the residential building, its electricity consumption is higher compared with the previous case. The nominal power of the required solar-PV system in order to zero its CO$_2$ emissions due to life cycle energy use is estimated from equation 7. Assuming that the C.O.P. of the heat pump is 3 and since X1=107.1 KWh/m^{2}year, X3= 20.4 KWh/m^{2}year, X4= 27.2 KWh/m^{2}year, Y=0.18 and Xemb = 30 KWh/m^{2}year, then:

P = 0.076 KWp/m^2 and P1= 0.062 KWp/m^2

Therefore, the nominal power of the necessary solar-PV system which will generate annually and inject into the grid the amount of electricity used for space heating with a heat pump, for lighting, for the operation of electric appliances as well as for annual compensation of the embodied energy of the building over a period of 50 years, is 0.062 KWp/m². For a residential building with a covered surface of 100 m² the nominal power is 6.2 KWp, which is higher than the power estimated in the previous case. Assuming that the solar-PV system will generate annually only the electricity used for the operation of the building (Xemb= 0), then equation 7A gives P1=0.046 KWp/m². For a residential building with a covered area at 100 m², the nominal power of the necessaey solar-PV system is 4.6 KWp. The results of the above-mentioned estimations are presented in table 3.

Table 3. Estimation of the required nominal power of a solar-PV system which could zero CO_2 emissions due to operating and life cycle energy use in a residential building in Crete, Greece with a covered area at 100 m² adjusted for grid electricity generation from non-CO_2 emitting fuels

Use of heating energy in the building	Space heating with solid biomass and hot water production with solar energy	Space heating with a heat pump with C.O.P.=3 and hot water production with solar energy
Nominal power of a solar-PV system generating annually the electricity used during its operation (KW_p)	2.6	4.6
Nominal power of a solar-PV system generating annually the electricity used during its operation plus its embodied energy (KW_p)	4.3	6.2
Additional power of the solar-PV system for covering its embodied energy (KW_p)	1.6	1.6
Cost of the additional solar-	2,400 €	2,400 €

PV system in the building for covering its embodied energy [1]		

[1] *Cost of the solar-PV system = 1,500 €/KWp*

Equations 6A, 6B and 5 indicate that the nominal power of the required solar-PV system in both cases depends on the share of the embodied energy to life cycle energy in the building, (a), as well as from the share of grid electricity generated from non-CO_2 emitting fuels, (Y). The power of the solar-PV system for different values of a is presented in table 4 while the power for different values of Y is presented in table 5.

Table 4. Nominal power of a solar-PV system which could zero CO_2 emissions due to life cycle energy use in a residential building in Crete, Greece (Y=0.18)

Share of the embodied energy to life cycle energy in the building	P1 (KWp/m²) [1]	P2 (KWp/m²) [2]
0.05	0.031	0.050
0.10	0.036	0.056
0.15	0.042	0.062
0.20	0.049	0.069
0.25	0.057	0.077
0.30	0.066	0.085

[1]*P1, Power of solar-PV when electricity is not consumed for space heating and for hot water production in the building,* [2]*P2, Power of solar-PV when a heat pump is used for space heating and electricity is not consumed for hot water production in the building*

Table 5. Nominal power of a solar-PV system which could zero CO_2 emissions due to life cycle energy use in a residential building in Crete, Greece (a=0.15)

Share of grid electricity generated by non-CO_2 emitting fuels	P1 (KWp/m²) [1]	P2 (KWp/m²) [2]
0.10	0.047	0.068
0.18	0.042	0.062
0.20	0.041	0.060
0.30	0.036	0.053
0.40	0.031	0.045
0.50	0.026	0.038

[1]*Power of solar-PV system when electricity is not consumed for space heating and hot water production,* [2]*Power of solar-PV system when a heat pump is used for space heating and electricity is not consumed for hot water production.*

6. Discussion

For creation of zero CO_2 emissions buildings due to energy use emphasis has been given in the use of renewable energy technologies. However, the improvement of building's energy efficiency reducing its overall energy consumption would result in lower annual energy requirements and lower size of the necessary renewable energy systems. Particularly important is the reduction of its heating and cooling needs which could be achieved with better thermal insulation. In the abovementioned analysis the grid electricity generated by renewable energies, including solar-PV and wind energy, which does not contribute in CO_2 emissions in Crete has been taken into account. Currently approximately 18 % of the total grid electricity is generated by renewable energies in the island. The recommended technologies for generation of heat, cooling and electricity in buildings are mature, reliable, well-proven and cost-effective. Solar-PVs are broadly used in southern European countries but their use in northern climate zones is rather limited. Therefore, offset of grid electricity use is more difficult in northern countries. Although separate use of the abovementioned renewable energy technologies is common in various buildings their combined use in order to zero their net CO_2 emissions due to energy use has not been reported as a high priority so far. Since these technologies are cost-effective financial incentives for their promotion are not necessary. Currently part of the European structural funds are utilized in Greece for the promotion of sustainable energy technologies in buildings, like solar thermal energy and solid biomass, in order to improve their energy performance and rating. The legal framework in Greece allows the use of solar-PVs in buildings in order to counterbalance the annual grid electricity use (operating electricity). However, in the case of installing a solar-PV system injecting more electricity than its annual consumption into the grid, additional financial compensation is not foreseen. Current European policies promote the creation of nearly zero energy buildings without mentioning or promoting zero carbon emissions buildings. The main barriers for the creation of zero CO_2 emissions buildings due to both operating and embodied energy use are related with:

a)   Lack of awareness among the public authorities and the citizens regarding the importance of this type of buildings,

b) Lack of a common accepted methodology for defining and estimating a net zero CO_2 emissions building,

c) Lack of pilot demonstration buildings with zero net CO_2 emissions used as good examples to the general public,

d) Lack of appropriate legal framework allowing offset and financial compensation of building's embodied energy generated on-site with solar-PV systems, and

e) Lack of appropriate European or national regulations promoting zero CO_2 emissions buildings like the existing regulations promoting nearly zero energy buildings.

7. Conclusions

Creation of net zero CO_2 emissions residential buildings due to life cycle energy use in Crete, Greece can be achieved without major difficulties. Life cycle energy consumption in a building is the sum of its operating and its embodied energy. In order to create such buildings, various reliable, mature and cost-effective renewable energy technologies can be used. With reference in the island of Crete, Greece, solar thermal energy, solar-PV energy, solid biomass and low enthalpy geothermal energy are available. The share of embodied energy varies in different type of buildings and on average it corresponds at 15 % of its life cycle energy use. It has been indicated that the embodied energy in a building can be offset with electricity generated with solar-PVs installed on-site and injected into the grid. A simple mathematical model has been developed for estimating the required solar-PV system, compensating its operating electricity as well as its embodied energy. The embodied energy in the building and the share of CO_2-free fuels used in grid electricity generation have been taken into account. Higher share of embodied energy to life cycle energy use in the building results in higher nominal power of the required solar-PV system. Higher share of grid electricity generated by non-CO_2 emitting fuels results in lower nominal power of the required solar-PV system. For a small residential building located in Crete, Greece with a covered area at 100 m^2, the required nominal power of the solar-PV system offsetting, additionally to its operating energy, its embodied energy has been estimated at 1.6 KWp and its cost at 2,400 €. However, the electricity injected into the grid offsetting its embodied energy is not financially compensated according to the current net-metering regulations in Greece. Creation of net zero CO_2 emissions buildings due to life cycle energy use has not been reported so far and future realization of such buildings requires removal of several barriers. Current work indicates that creation of net zero CO_2 emissions buildings due to life cycle energy use can be achieved in the future in a cost effective way using various existing and well-proven renewable energy technologies promoting energy sustainability.

References

[1] Torcellini, P., Pless, S., Deru, M. & Crawley, D. (2006). "Zero energy buildings: A critical look at the definition", National laboratory of the US Department of Energy, presented at *ACEEE summer study, Pacific Grove, California*, August, 14-18, 2006.

[2] Marszal, A. J., Heiselberg, P., Bourrelle, J. S., Musall, E., Voss, K., Sartori, I. & Napolitano, A. (2011). "Zero Energy Building – A review of definitions and calculation methodologies", *Energy and Buildings*, 43(4), pp. 971–979. doi:10.1016/j.enbuild.2010.12.022

[3] Vourdoubas, J. (2016). "Creation of zero CO_2 emissions residential buildings due to energy use, A case study in Crete, Greece", *Journal of Civil Engineering and Architecture Research*, 3, pp. 1251-1259.

[4] Vourdoubas, J. (2017). "Realization of a small residential building with zero CO_2 emissions due to energy use in Crete, Greece", *Studies in Engineering and Technology*, 4, pp. 112-120, doi:10.11114/set.v4i1.2567

[5] Tselepis, S. (2015). "The PV market development in Greece, Net-metering study cases". Retrieved on 19/9/2017 from http://www.cres.gr/kape/publications/photovol/new/S

[6] Ibn-Mohammed, T., Greenough, R., Taylor, S., Ozawa-Meida, L. & Acquaye, A. (2013). "Operational vs. embodied emissions in buildings—A review of current trends", *Energy and Buildings*, 66, pp. 232–245. doi:10.1016/j.enbuild.2013.07.026

[7] Suzuki, M., Oka, T. & Okada, K. (1995). "The estimation of energy consumption and CO_2 emission due to housing construction in Japan", *Energy and Buildings*, 22(2), pp. 165–169. doi:10.1016/0378-7788(95)00914-j

[8] Cuéllar-Franca, R. M. & Azapagic, A. (2012). "Environmental impacts of the UK residential sector: Life cycle assessment of houses", *Building and Environment*, 54, pp. 86–99. doi:10.1016/j.buildenv.2012.02.005

[9] Syngros, G., Balaras, C. A. & Koubogiannis, D. G. (2017). "Embodied CO_2 Emissions in Building Construction Materials of Hellenic Dwellings", *Procedia Environmental Sciences*, 38, pp. 500–508. doi:10.1016/j.proenv.2017.03.113

[10] Karimpour, M., Belusko, M., Xing, K. & Bruno, F. (2014). "Minimizing the life cycle energy of buildings: Review and analysis", *Building and Environment*, 73, pp. 106–114. doi:10.1016/j.buildenv.2013.11.019

[11] Dixit, M. K., Fernández-Solís, J. L., Lavy, S. & Culp, C. H. (2012). "Need for an embodied energy measurement protocol for buildings: A review paper", *Renewable and Sustainable Energy Reviews*, 16(6), pp. 3730–3743. doi:10.1016/j.rser.2012.03.021

[12] Cellura, M., Guarino, F., Longo, S. & Mistretta, M. (2014). "Energy life-cycle approach in net zero energy buildings balance: Operation and embodied energy of an Italian case study", *Energy and Buildings*, 72, pp. 371–381. doi:10.1016/j.enbuild.2013.12.046

[13] Dixit, M. K., Fernández-Solís, J. L., Lavy, S. & Culp, C. H. (2010). "Identification of parameters for embodied energy measurement: A literature review", *Energy and Buildings*, 42(8), pp. 1238–1247. doi:10.1016/j.enbuild.2010.02.016

[14] Berggren, B., Hall, M. & Wall, M. (2013). "LCE analysis of buildings – Taking the step towards Net Zero Energy Buildings", *Energy and Buildings*, 62, pp. 381–391. doi:10.1016/j.enbuild.2013.02.063

[15] Hernandez, P. & Kenny, P. (2011). "Development of a methodology for life cycle building energy ratings", *Energy Policy*, 39(6), pp. 3779–3788. doi:10.1016/j.enpol.2011.04.006

[16] Sartori, I. & Hestnes, A. G. (2007). "Energy use in the life cycle of conventional and low-energy buildings: A review article", *Energy and Buildings*, 39(3), pp. 249–257. doi:10.1016/j.enbuild.2006.07.001

[17] Ramesh, T., Prakash, R. & Shukla, K. K. (2010). "Life cycle energy analysis of buildings: An overview", *Energy and Buildings*, 42(10), pp. 1592–1600. doi:10.1016/j.enbuild.2010.05.007

[18] Adalberth, K., Almgren, A. & Petersen, E.H. (2001). "Life cycle assessment of four multi-family buildings", *International Journal of Low Energy and Sustainable Buildings*, 2, pp. 1-21.

[19] Scheuer, C., Keoleian, G. A. & Reppe, P. (2003). "Life cycle energy and environmental performance of a new university building: modeling challenges and design implications", *Energy and Buildings*, 35(10), pp. 1049–1064. doi:10.1016/s0378-7788(03)00066-5

[20] Ramesh, T., Prakash, R. & Kumar Shukla, K. (2013). "Life Cycle Energy Analysis of a Multifamily Residential House: A Case Study in Indian Context", *Open Journal of Energy Efficiency*, 2, pp. 34–41. doi:10.4236/ojee.2013.21006

[21] Fay, R., Treloar, G. & Iyer-Raniga, U. (2000). "Life-cycle energy analysis of buildings: a case study", *Building Research & Information,* 28(1), pp. 31–41. doi:10.1080/096132100369073

[4] Use of renewable energies for creation of net zero carbon emissions residential buildings in northern Greece

1. Introduction

Increase of energy efficiency in buildings and decrease of their carbon emissions due to energy use are important measures for climate change mitigation. Various renewable energies can be used in residential buildings for covering their needs in air-conditioning, electricity and domestic hot water. Current advances in green energy technologies allow their use in a cost-effective way in residential buildings as well as in other applications in different sectors. Current research aims in the investigation of the possibility of using renewable energies for covering all the energy requirements in residential buildings located in Northern Greece. It is indicated that locally available renewable energies could provide heat, cooling and electricity in these buildings zeroing their needs for fossil fuels and grid electricity based on them. Present research is important and useful because it indicates the way that various mature, reliable and cost-effective renewable energy technologies can be used for energy generation in residential buildings covering all their energy requirements and zeroing their carbon footprint due to energy use. This complies with National and EU goals for reduction of GHG emissions while it offers many economic and social advantages in local societies.

2. Literature survey

2.1 Use of solid biomass in individual heating systems

Carlini et al, 2013 have reported on the economic assessment of biomass boiler plants for heating Italian residential buildings. The authors stated that solid biomass, including wood and pellet boilers, can provide heat for space heating and domestic hot water production in a house. They concluded that installation of a biomass boiler results in economic benefits which would be higher if a governmental subsidy Is offered. A review of biomass heating in UK homes has been presented by Dwyer, 2006. The author stated that use of biomass for heating should be considered in the early stages of the building's design. A study of biomass boilers' heat generation in residential buildings in Spain has been reported by Las-Heras-Casas et al, 2018. The authors stated that biomass use in the residential sector can assist in the achievement of EU's 2020 goals for climate and energy. They also stated that biomass use can reduce primary non-renewable energy consumption in buildings by 93% while CO_2 emissions by 94%.

2.2 Use of solid biomass in district heating systems

Ericsson et al, 2016 have reported on introduction and expansion of biomass use in Swedish district heating systems. The authors stated that district heating satisfies about 60% of heat demand in Swedish buildings and biomass alone accounts for about half of heat supply. Design of biomass district heating systems has been reported by Vallios et al, 2007. The authors presented a methodology for designing biomass-fuelled district heating systems using a parametric logic which assists an inexperienced engineer to be able to design such systems. Biomass district heating in pilot installations for public buildings has been mentioned by Chatzistougianni et al, 2016. The authors proposed a methodology for heating public buildings with district heating systems fuelled by biomass. They mentioned that benefits for rural municipalities are important in terms of operating costs and environmental protection. Margaritis et al, 2014 have reported on the introduction of renewable energies in district heating systems in Greece. The authors stated that various district heating systems in northern Greece are utilizing waste heat rejected from lignite-fuelled thermal power stations. However, these power stations are going to close down in coming years. They have also estimated that locally available solid biomass can be used instead of waste heat in existing district heating systems while fuel replacement is cost-effective.

2.3 Use of waste heat in district heating systems

Dorasic et al, 2018 have reported on excess heat utilization in district heating systems. The authors stated that district heating systems offer many economic and environmental benefits compared with individual heating systems. They also reported that excess heat can be easily utilized in a district heating system depending on available excess heat supply and its distance from the area of heat consumption. Karlopoulos et al, 2004 have reported on the experience of district heating systems operating in northern Greece using waste heat rejected from lignite-fired thermal power plants. The authors stated that many economic and environmental benefits have resulted during the last ten years due to the operation of these systems in northern Greece. Fang et al, 2013 have investigated industrial waste heat utilization for low-temperature district heating. The authors proposed a holistic approach for integrated and efficient utilization of low grade industrial waste heat. They mentioned four important advantages including: a) Improvement of thermal energy efficiency in factories, b) More cost-efficient heating than traditional methods, c) Reduction of GHG emissions, and d) Reduction of heat pollution due to industrial waste heat discharge.

2.4 Use of solar thermal energy for domestic hot water production

Greening et al, 2014 have reported on domestic solar water heating in United Kingdom. Taking into account the low solar irradiance in U.K. the authors stated that the potential of solar thermal systems to contribute in a more sustainable domestic energy supply in U.K. is limited. A review of solar water heating systems has been presented by Vinubhai et al, 2014. The authors stated that solar water heating is one of the most effective technologies to convert solar energy into thermal energy and it is currently a well developed and commercialized technology. However, they mentioned, opportunities exist for further improvement of system's performance and increase of their reliability and efficiency. Aelenei et al, 2016 have reported on systematic characterization of solar thermal systems installed in buildings. The authors stated that apart from their energy performance characterization additional criteria for assessing these systems should include structural, functional and aesthetical aspects. Tian et al, 2013 have reviewed solar collectors and thermal energy storage in solar thermal applications. The authors reviewed various types of concentrated and non-concentrated solar collectors with or without sun-tracking systems. They also reviewed solar thermal energy storage as sensible heat, latent heat and in heat storage materials. The authors mentioned that among non-concentrated solar collectors the photovoltaic-thermal collectors have shown the best overall performance. Kalogirou, 2009 has reported on thermal performance, economic and environmental life cycle analysis of thermosiphonic solar water heaters. The author studied a simple thermosiphonic system producing hot water for a family of four persons. His results indicated that it can cover up to 79% of the annual needs in domestic hot water of the family. The payback period of the investment varies between 2,7-4,5 years, depending on the back-up system used, while GHG savings obtained are at 70%.

2.5 Use of solar-PV systems

Salem et al, 2015 have reported on building –integrated photovoltaics in Mediterranean countries. The authors stated that Mediterranean basin is characterized by high solar irradiance, at 7.5- 8 KWh/m² annually, and high temperatures which decrease the efficiency of the solar cells. They concluded that solar-PVs placed on commercial buildings, which mainly operate during the day, can cover most of their electricity needs. Hayter et al, 2002 have reported on applications of solar-PVs in buildings. The authors investigated the performance of three solar-PV systems, at 7 KWp to 60 KWp, installed in three commercial buildings. They stated that these systems reduced the electricity loads in buildings, suggesting that they should be studied and integrated in its initial design phase. Biyik et al, 2017 have reported on building-integrated photovoltaic systems (BIPV). The authors stated that BIPVs are considered as a feasible technology to cover part of the electricity load in buildings. They also investigated the possibility of ventilating and cooling the solar-PV system in order to decrease the temperature of the panels and to increase their efficiency. Tselepis, 2015 has reported on solar-PV market developments in Greece with reference to net-metering case studies. The author stated that net metering regulations were introduced in Greece at the end of 2014. The two case studies presented for a household and a commercial enterprise indicated their attractiveness and the viability of the net-metering program in Greece.

2.6 Net zero energy and net zero carbon emission buildings

Ferrante et al, 2011 have investigated the creation of net zero energy balance and net zero on-site CO_2 emissions houses in Mediterranean climate. With reference to a town in southern Italy the authors stated that houses with net zero energy balance and net zero CO_2 emissions are feasible in Mediterranean basin. They can be achieved using energy-efficient technologies, local materials and traditional construction processes. Vourdoubas, 2016 has reported on the creation of net zero CO_2 emissions residential buildings in the island of Crete, Greece. The author stated that zero carbon emissions buildings can be created with the combination of various renewable energy technologies. He stated that the combined use of solar thermal energy, solar-PV, solid biomass and ground source heat pumps can generate all the heat and electricity required in the building, zeroing its net carbon footprint due to energy use. Creation of net zero CO_2 emissions residential buildings due to operating and embodied energy use has been reported by Vourdoubas, 2017. The author stated that combination of various renewable energies could zero their CO_2 emissions due to operating energy use. He also mentioned that in order to zero the embodied energy in the building, generation of additional solar electricity injected into the grid is required. Nielsen et al, 2012 have reported on excess heat production in net zero energy buildings (NZEBs) within district heating areas in Denmark. The authors stated that most buildings in Denmark are connected to electricity grids and around half of them with district heating systems. This fact allows exchange of heat and electricity among buildings and grids. They mentioned that excess heat produced from solar thermal systems installed in NZEBs can be injected into the district heating system reducing the fuel used for heat generation.

The aims of the current work are:

a)   A presentation of various sustainable energy technologies which can be used in residential buildings in northern Greece for zeroing their net carbon emissions,

b)   A preliminary sizing of sustainable energy technologies used in a typical residential building for zeroing its net carbon footprint, and

c)   An assessment of the feasibility of creating net zero carbon emissions residential buildings located in northern Greece.

3. Energy use in residential buildings

Residential buildings consume energy in various sectors including:
a)   Space heating,
b)   Space cooling,
c)   Domestic hot water production,
d)   Lighting, and
e)   Operation of various electric appliances

Energy consumption in various sectors in residential buildings depends on various parameters including:
a)   The local climate,
b)   The quality of the building construction which characterizes its energy performance, and
c)   The behavior of the residents

Typical operating energy use in various sectors in a residential building located in Crete, Greece is presented in Table 1.

Table 1. Energy consumption in a residential building located in Crete, Greece [1]

Sector	% energy used	Annual energy consumption (KWh/m²)
Space heating	63	107.1
Hot water production	9	15.3
Lighting	12	20.4

| Operation of various electric appliances including those for space cooling | 16 | 27.2 |
| Total | 100 | 170 |

[1]Source: Vourdoubas, 2016

4. Requirements for net zero carbon emission buildings due to energy use

A grid-connected residential building could zero its net carbon emissions due to operating energy use if the following conditions are fulfilled:

a) All its heating requirements for space heating and hot water production are covered with renewable energies or other non-carbon emitted energy resources, instead of fossil fuels, and

b) Its grid electricity consumption is offset annually with green electricity, like solar-PV electricity generated by solar-PV panels placed on the building.

If these two conditions are fulfilled the net carbon emissions due to operating energy use in the residential building would be zero. It has been assumed though that all grid electricity is generated by fossil fuels emitting CO_2 in the atmosphere. However, part of grid electricity in Greece is generated by renewable energies including hydroelectric energy, wind energy, biogas and solar-PV plants. Apart from the energy consumed during the operation of the building, additional energy is consumed during its construction, its refurbishment and its demolition. Various studies indicate that operating energy use in a building corresponds approximately at 85% of its total life cycle energy consumption while the remaining 15% is its embodied energy. In order to zero its net carbon emissions due to life cycle energy consumption, a larger size solar-PV system is required. However, the existing net-metering regulations in Greece compensate only the solar electricity generated which is equal to the annual consumption in the building. Excess electricity generated is injected into the grid while the producer-consumer does not receive any financial compensation for that.

5. Availability of solar energy, solid biomass and waste heat in northern Greece

Solar energy is abundant in northern Greece and it is already used for heat and power generation with solar thermal and solar-PV systems. Both of them are currently used in buildings for covering their energy needs. The green energy systems used are mature, reliable and cost-effective. Use of solar-PV systems for electricity generation in grid-connected buildings is allowed according to Greek legislation with net-metering regulations. Solar thermal systems are broadly used in Greece for domestic hot water production. However, their use for space heating and cooling is rather limited. Availability of solid biomass in northern Greece is high, including agricultural and forest by-products, residues and wastes. It is broadly used for heat generation, including space heating and domestic hot water production in residential buildings. Various types of solid biomass are used for energy generation including firewood, pellets and wood briquettes feeding wood stoves, fireplaces and central heating systems. Although use of solid biomass in residential buildings in northern Greece is well developed its use in fuelling district heating systems is lacking. Taking into account the current severe economic crisis in the country, heating residential buildings with solid biomass instead of conventional fuels is considered as a cheap and affordable method of heating. Availability of waste heat in northern Greece is high, particularly in areas with lignite-fired thermal power stations. These power plants discharge large quantities of hot water which could be recycled and reused in district heating systems providing heat in buildings and in other activities. Although waste heat from these power stations cannot be considered as a renewable energy source, its reuse is desirable promoting the circular economy, increasing the overall efficiency in power stations, avoiding thermal pollution of water reservoirs and eliminating carbon emissions due to fossil fuels use. Currently district heating systems utilizing waste heat discharged from power stations provide heating in few cities in northern Greece including Kozani and Ptolemaida which are located nearby the power stations. Solar energy, solid biomass and waste heat, if combined, can cover all the energy needs in residential buildings, resulting in zero net carbon emissions due to operating energy use. The use of solar energy, solid biomass and waste heat for covering the energy needs in residential buildings is presented in Table 2.

Table 2. Use of solar energy, solid biomass and waste heat for covering the energy needs in residential buildings [1]

Energy source	Space heating	Domestic hot water production	Electricity generation
Solar thermal energy		+	
Solar-PV energy			+

| Solid biomass in individual or district heating systems | + | + | |
| Waste heat in district heating systems | + | + | |

[1] Source: Own estimations

6. Use of solid biomass and solar energy for covering all the energy needs in residential buildings

Renewable energies can cover all the heating needs in residential buildings while solar-PVs placed on their roofs can generate electricity equal to the amount consumed annually in the building. Solid biomass can be used in individual systems for space heating and hot water production while solar thermal systems with flat plate collectors can be used for hot water production. Both technologies are reliable, mature and cost-effective and they are currently used in residential buildings in northern Greece. Solar-PV systems are used in grid-connected buildings for electricity generation, offsetting annually its grid electricity consumption according to net-metering regulations. Their use has been promoted in the last few years due to the fact that their prices have been sharply reduced and their use is profitable. Alternatively district heating systems fuelled with solid biomass can provide all the required heat in the buildings. Unfortunately district heating systems utilizing solid biomass are not currently operating in Greece. Therefore, the combined use of solid biomass, solar thermal energy and solar-PV systems can cover all the energy requirements in grid-connected residential buildings resulting in net zero carbon emissions due to operating energy use in them. The characteristics of the above-mentioned renewable energy technologies are presented in Table 3.

Table 3. Characteristics of various renewable energy technologies and waste heat used for energy generation in residential buildings in northern Greece [1]

Energy source/ technology	Energy generation	Energy use in buildings	Reliability of technology	Cost effectiveness of technology	Current use in buildings in northern Greece
Solid biomass in individual systems	Heat	Space heating, Hot water production	High	Satisfactory	Yes
Solid biomass in district heating systems	Heat	Space heating, Hot water production	High	Satisfactory	No
Solar thermal energy-flat plate collectors	Heat	Hot water production	High	Satisfactory	Yes
Solar energy-PV systems	Electricity	Lighting, operation of electric appliances	High	Satisfactory	Yes
Waste heat rejected from various plants	Heat	Space heating, Hot water production	High	Satisfactory	Yes

[1] Source: Own estimations

Sizing the necessary renewable energy systems providing all the required energy in a residential building located in northern Greece has been calculated according to the following assumptions:

a)   The covered area of the house is 150 m^2,

b) Energy consumption in various sectors in the house is similar with that presented in table 1. Energy requirements for space heating are 14,175 KWh$_{th}$/year, for hot water production 2,025 KWh$_{th}$/year and for electricity 6,300 KWh$_{el}$/year, totally 22,500 KWh/year,

c) Solid biomass is used in individual systems covering all its needs in space heating and 50% of its needs in domestic hot water. Solid biomass should provide 15,187.5 KWh$_{th}$/year,

d) A solar thermal system covers the remaining 50% of its needs in domestic hot water providing 1,012.5 KWh$_{th}$/year. A system with flat plate collectors at 2 m^2 can provide it,

e) A solar-PV system installed on building's roof and connected with the electric grid generates annually the amount of grid electricity consumed in it according to net-metering initiative generating 6,300 KWh$_{el}$/year. It is assumed that a solar-PV system located in northern Greece generates annually 1,400 KWh/KW$_p$.

The required renewable energy systems in the residential building are:

a) A solid biomass burning system with thermal power at 15 KW$_{th}$,
b) A solar thermal system with flat plate collectors at 2 m^2, and
c) A solar-PV system with nominal power at 4.5 KW$_p$.

The cost of the above-mentioned green energy systems is estimated assuming that their unit costs are: for the biomass burning system 300 €/KW$_{th}$, for the solar thermal system 400 €/m^2 of the solar collector and for the solar-PV system 1,200 €/KW$_p$. Therefore, the cost of the biomass burning system is at 4,500 €, of the solar thermal system at 800 € and of the solar-PV at 5,400 €. The overall cost is 10,700 € or 71.33 €/m^2 of the building's covered surface.

7. Use of waste heat and solar energy for covering all the energy needs in residential buildings

Waste heat rejected from thermal power plants in northern Greece is currently used in district heating systems providing heat in residential buildings. Although rejected waste heat cannot be considered as a renewable energy source its impacts are similar with renewable energies since its use is not related with GHG emissions while thermal pollution to reservoirs is avoided. Additionally, pricing of rejected waste heat is very low and the cost of heating the residential buildings is lower compared with a district heating system fuelled with solid biomass or conventional fuels. District heating systems using waste heat can also utilize solid biomass or heat produced from incineration of various wastes. They can also utilize heat generated by solar thermal systems. A solar-PV system can provide all the electricity consumed annually in the house. In this case the necessary renewable energy systems in the residential building include only the solar thermal system and the solar-PV system while space heating will be provided by the district heating system:

a)   The district heating system will provide annually 15,187.5 KWh_{th}/year,
b)   A solar thermal system with flat plate collectors at 2m² will provide 1,012.5 KWh_{th}/year, and
c)   A solar-PV system with nominal power 4.5 KWp will provide 6,300 KWh_{el}/year.

The cost of these energy systems is estimated similarly as in section 6. In this case an individual space heating system is not needed and the cost of both the solar thermal and the solar-PV system is 6,200 € or 41.33 €/m² of the building's covered surface. The capital cost of the sustainable energy systems providing all the required energy in the residential building is presented in Table 4.

Table 4. Capital cost of sustainable energy systems covering all the energy needs of a house with a covered area of 150 m² located in northern Greece[1]

Energy system	Use of a solid biomass burning system, a solar thermal system and a solar-PV system (€)	Use of a district heating system fuelled with rejected waste heat, a solar thermal system and a solar-PV system (€)
Solid biomass burning system	4,500	-
Solar thermal system	800	800
Solar-PV system	5,400	5,400
Total	10,700	6,200
Total costs per m² of covered area	71.33	41.33

[1]Unit costs: Solid biomass burning system, 300 €/KW_{th}, Solar thermal system, 400 €/m² of flat plate collectors, Solar-PV system, 1,200 €/KW_p

8. Discussion

Residential buildings in few towns in northern Greece can be connected in two energy grids including the electric grid, with net-metering regulations and the municipal district heating grid fuelled by industrial waste heat. The building's connection with the two energy grids allows the electricity generation with photovoltaic panels and its injection into the grid. The building can also generate heat with solar thermal systems installed on its roof and injected into the district heating network when it is not needed. Therefore, building's owners are becoming "pro-sumers" producing and consuming electric and thermal energy. Apart from using solid biomass and waste heat in the residential building another alternative for its space heating is the use of high efficiency heat pumps, a reliable and cost effective technology. However, heat pumps use electricity and their initial cost is high. If they are used the necessary solar-PV system should be of higher capacity for generating the additional electricity consumed by them. The use of energy-saving techniques and technologies reducing its overall energy consumption has not been considered so far and emphasis has been focused only in zeroing fossil fuels use and CO_2 emissions. Reducing energy consumption in the residential building will result in smaller sizing of the required benign energy systems and in lower capital investments for zeroing its carbon emissions due to energy use. However, according to current EU directives and the Greek legislation complying with these directives, reduction of electric and thermal energy consumption is of high priority. In the case that space heating in the building is provided by the district heating system the capital cost of the necessary renewable energy systems is lower compared with the cost in the case of using solid biomass for heat production. However, in both cases the capital cost of green energy systems is low compared with the construction cost of the building which exceeds 1,000 €/m² of the building's covered surface.

9. Conclusions

The combined use of various renewable and non-carbon emitted energy resources could result in zeroing carbon emissions in residential buildings in northern Greece. Due to high availability of various sustainable energies in northern Greece creation of net zero carbon emissions residential buildings is feasible. Solar energy, solid biomass and waste heat rejected from thermal power plants operating in the region can be used for that. Their technologies for heat and power generation are mature, reliable and cost-effective. The legal framework allows their use in residential buildings for heat and electricity generation. Two different options have been studied. In the first solar thermal energy, solar-PV energy and solid biomass are used for covering all energy needs in the building. In the second solar thermal energy, solar-PV energy and industrial waste heat are used for that. The capital cost of the abovementioned renewable energy systems zeroing the net carbon footprint due to energy use in the house varies between 6,200 € to 10,700 € or 41.33 €/m^2 to 71.33 €/m^2 of the building's covered surface. Current work indicates that creation of net zero CO_2 emissions buildings due to operating energy use in northern Greece using locally available sustainable energies is technically feasible and economically attractive. Further work should be focused on the realization of a net zero carbon emissions residential building in northern Greece using the above-mentioned sustainable energy systems. Study of its energy behavior will indicate its energy neutrality.

References

[1] Aelenei, L., Smyth, M., Platzer, W., Norton, B., Kennedy, D., Kalogirou, S. & Maurer, Ch. (2016). "Solar thermal systems - Towards a systematic characterization of building integration", *Energy Procedia*, 91, pp. 897-906. doi: 10.1016/j.egypro.2016.06.256

[2] Biyik, E., Araz, M., Hepbasli, A., Shahrestani, M., Yao, R., Shao, L., Essah, E., Oliveira, C.A., Del Cano, T., Rico. E., Lechon, J.L., Andrade, L., Mendes, A. & Atli, Y.B. (2017). "A key review of building integrated photovoltaic (BIPV) systems", *Engineering Science and Technology, An International Journal,* 20, pp. 833-858. https://doi.org/10.1016/j.jestch.2017.01.009

[3] Carlini, M., Castellucci, S., Cocchi, S., Allegrini, E. & Li, M. (2013). "Italian residential buildings: Economic assessment for biomass boilers plants", *Mathematical Problems in Engineering*, Article 823851, http://dx.doi.org/10.1155/2013/823851

[4] Chatzistougianni, N., Giagozoglou, E., Sentzas, K., katastergios, E., Tsiamitros, D., Stimoniaris, D., Stomoniaris, A. & Maropoulos, S. (2016). "Biomass district heating methodology and pilot installations for public buildings groups", in *20th Innovative Manufacturing Engineering and Energy Conference.* doi:10.1088/1757-899X/161/1/012083

[5] Dorasic, B., Novosel, T., Puksec, T. & Duic, N. (2018). "Evaluation of excess heat utilization in district heating systems by implementing levelized cost of excess heat", *Energies*, 11, 575, doi:10.3390/en11030575

[6] Dwyer, T. *(2006)*. *"A review of biomass heating for UK homes and commercial applications"*, *International Journal of low Carbon Technologies,* 1(4), pp. 329-335. https://doi.org/10.1093/ijlct/1.4.329

[7] Ericsson, K. & Werner, S. (2016). "The introduction and expansion of biomass use in Swedish district heating systems", *Biomass and Bioenergy*, 94, pp. 57-65, http://dx.doi.org/10.1016/j.biombioe.2016.08.011

[8] Fang,H., Xia, J., Zhu, K., Su, Y. & Jiang, Y. (2013). "Industrial waste heat utilization for low temperature district heating", *Energy Policy*, 62, pp. 236-246. http://dx.doi.org/10.1016/j.enpol.2013.06.104

[9] Ferrante, A. & Cascella, M.T. (2011). "Zero energy balance and zero on-site CO_2 emission housing development in Mediterranean climate", *Energy and Buildings*, 43, pp. 2002-2010. doi:10.1016/j.enbuild.2011.04.008

[10] Greening, B. & Azapagic, A. (2014). "Domestic solar thermal water heating: A sustainable option for the UK?", Rene*wable Energy*, 63, pp. 23-36. http://dx.doi.org/10.1016/j.renene.2013.07.048

[11] Hayter, S., Torcellini, P. & Deru, M. (2002). "Photovoltaics for buildings: New applications and lessons learnt", *American council for an energy efficient economy, Summer study on energy efficiency in buildings,* Pacific Grove, California, USA, August 18023. 2002. Retrieved at 4/6/2018 from https://www.nrel.gov/docs/fy02osti/32158.pdf

[12] Kologirou, S. (2009). "Thermal performance, economic and environmental life cycle analysis of thermosiphonic solar water heaters", *Solar Energy*, 83, pp. 39-48. doi:10.1016/j.solener.2008.06.005

[13] Karlopoulos, E., Pekopoulos, D. & Kakaras, E. (2004). "District heating systems from lignite-fired power plants—Ten years experience in Greece". In Proceedings of the *International Workshop on Promotion of CHP/Tri-generation and Collaborative Potentials in Chinese Market,* Hangzhou, China, 26–28 April 2004.

[14] Las-Heras-Casas, J., Lopez-Ochoa, L.M, Paredes-Sanchez, J.P. & Lopez-Gonzalez, L.M. (2018). "Implementation of biomass boilers for heating and hot water in multi-family buildings in Spain: Energy, environmental and economic assessment", *Journal of Cleaner Production*, 176, pp. 590-603. https://doi.org/10.1016/j.jclepro.2017.12.061

[15] Margaritis, N., Rakopoulos, D., Mylona, E. & Grammelis, P. (2014). "Introduction of renewable energy sources in the district heating system of Greece", *International Journal of Sustainable Energy Planning and Management*, 4, pp. 43-56. dx.doi.org/10.5278/ijsepm.2014.4.5

[16] Nielsen, S. & Moller, B. (2012). "Excess heat production of future net zero energy buildings within district heating areas in Denmark", *Energy*, 48, pp. 23-31. doi:10.1016/j.energy.2012.04.012

[17] Salem, T. & Kinab, E. (2015). "Analysis of building-integrated photovoltaic systems: A case study of commercial buildings under Mediterranean climate", *Procedia Engineering*, 118, 5380545. doi: 10.1016/j.proeng.2015.08.473

[18] Tian, Y. & Zhao, C.Y. (2013). "A review of solar collectors and thermal energy storage in solar thermal applications", *Applied Energy*, 104, pp. 538-553. http://dx.doi.org/10.1016/j.apenergy.2012.11.051

[19] Tselepis S. (2015). "The PV Market Developments in Greece, Net-Metering Study Cases". Retrieved at 25/6/2018 from http://www.cres.gr/kape/publications/photovol/new/S%20%20Tselepis%20%20The%20PV%20Market%20Developments%20in%20Greece%20%20Net-Metering%20Study%20Cases%2031st%20EUPVSEC%202015%20Hamburg%20%207DV.4.26.pdf

[20] Valios, I. & Tsoutsos, Th. (2007). "Design of biomass district heating systems", In *15th European Biomass Conference*, Berlin, Germany, 7-11 May 2007.

[21] Vinubhai, T.S., Vishal, R.J. & Thakkar, K. (2014). "A review: Solar water heating systems", in the *National conference on emerging Vista of technology in the 21st century*. doi: 10.13140/2.1.1910.5281

[22] Vourdoubas, J. (2016). "Creation of zero CO_2 emission residential building due to energy use: A case study in Crete, Greece", *Journal of Civil Engineering and Architecture Research*, 3(2), pp. 1251-1259.

[23] Vourdoubas, J. (2017). "Creation of Zero CO_2 Emissions Residential Buildings due to Operating and Embodied Energy Use on the Island of Crete, Greece", *Open Journal of Energy Efficiency*, 6, pp. 141-154. https://doi.org/10.4236/ojee.2017.64011

[5] Review of sustainable energy technologies use in buildings in Mediterranean basin

1. Introduction

Current policies in various countries try to mitigate climate change which consists of one major global environmental threat increasing the use of renewable energies replacing fossil fuels. Technological improvements of renewable energy technologies have increased their reliability and their cost-effectiveness in heat and power generation. Buildings consume large amounts of energy during their operation emitting large amounts of GHGs into the atmosphere. Use of various renewable energy technologies covering their energy requirements in heat and electricity is easier compared with their use in industry, agriculture and transportation. European directives require that new buildings, constructed after 2020, should be nearly zero energy buildings reducing their energy consumption and carbon emissions. Creation of net zero energy buildings with zero CO_2 emissions, due to energy use is an interesting challenge. Mediterranean basin has abundant renewable energy resources which can be used for heat and power generation covering a large part of energy needs in many countries located in this geographical area. Review of sustainable energy technologies used in buildings in Mediterranean basin is important since it indicates the benign green energy technologies which can increase energy sustainability and performance in buildings reducing or zeroing their carbon footprint due to energy use.

2. Literature survey

2.1 Solar thermal systems with flat plate collectors

A study concerning the advantages and drawbacks of domestic solar heating systems in UK has been implemented by Greening et al, 2014. The authors compared solar thermal systems with gas boilers, electricity and heat pumps according to eleven (11) environmental criteria. They stated that they have many environmental advantages but due to their poor efficiency and the need for a back-up system their future development in UK seems rather limited. A review of solar water heating systems has been presented by Vinubhai et al, 2014. The authors stated that solar water heating is one of the most effective technologies to convert solar energy into thermal energy and it is already a well developed and commercialized technology. However, they concluded, there are opportunities for further improvements in the reliability and efficiency of these systems. A review regarding characterization of building-integrated solar thermal systems has been presented by Aelenei et al, 2016. The authors mentioned that in the past the decision for installing a solar thermal system in a building was based on techno-economic evaluations. Currently the decision is based on additional criteria including architectural integration, aesthetics, functional and environmental characteristics.

2.2 Solar parabolic collectors

Design of a solar parabolic dish with a Sterling engine has been presented by Hafez et al, 2016. The authors studied, with simulation techniques, the influence of various design parameters on performance of the system. Depending on reflector's materials the power efficiency of the solar parabolic dish has been estimated between 49.52% and 97.07%. Quintal et al, 2011 have reported on the use of parabolic trough solar collectors for air conditioning and hot water production in buildings in Portugal. The authors stated that a small size system (10 KW) combined with absorption technology can be roof-mounted in a building and it can provide both air conditioning and hot water to it. They have also estimated that the system is cost-effective, having a payback period of eight (8) years without governmental subsidies.

2.3 Solar thermal cooling

A case study for a solar cooling system in a public building in Rome, Italy has been presented by Grignaffini et al, 2012. The authors stated that during the summer the demand for electricity increases in Mediterranean countries due to cooling requirements in buildings. The use of solar thermal energy with absorption chillers for space cooling is attractive since solar energy is abundant when cooling is needed. They estimated that the solar cooling system can reduce the cooling load in the building by 26.5%. A report on solar air conditioning has been published by International Energy Agency, 2011. According to this report solar absorption technology is at a critical stage. The technology has shown that significant energy savings are possible and it has reached a level of early market deployment. However, the financial risks of using this technology are still too high. A review of solar cooling technologies for residential applications in Canada has been reported by Baldwin et al, 2012. The authors stated that use of solar cooling systems can assist in the reduction of energy consumption for space cooling in buildings. They also stated that limited work has been conducted so far in the area of small-scale systems worldwide while most of existing solar cooling applications are related with large buildings. Use of solar heating and cooling in buildings has been reported by Henning et al, 2012. The authors presented a design study of a solar thermal system providing heat and cooling in a hotel located in Malta in Mediterranean basin. Their results indicated that the life cycle cost of a green solar heating and cooling system was not higher than a conventional system while primary energy savings up to 80% can be obtained.

2.4 Solar photovoltaic use

An analysis of building-integrated solar photovoltaics (solar-PVs) in Mediterranean climate has been presented by Salem et al, 2015. The authors have conducted a literature review regarding building-integrated photovoltaics (BIPVs). They stated that they can produce an adequate amount of energy while being at the same time part of the building envelope. The use of solar-PVs in buildings has been studied by Hayter et al, 2002. The authors have investigated the performance of solar-PVs integrated in three commercial buildings in USA. They mentioned that use of solar-PVs in buildings offers many benefits and their use should be a standard consideration when designing a building. A review of building-integrated PVs has been presented by Biyik et al, 2017. The authors have investigated and compared the use of single BIPVs together with the use of hybrid BIPVs generating heat and power for covering both electric and thermal loads. They stated that existing technologies could be improved by a) ventilating the solar panels lowering their temperature and increasing their yield, and b) by using thin film technologies integrating better the solar panels into the building envelope.

2.5 Wind turbine use

Building-integrated wind turbines have been studied by Bobrova, 2015. The author stated that perhaps in the future the wind generator will become an integral part of houses. Then, she stated, the form, structure and volume of the building will depend on integrating wind generators to them. Urban wind turbine integration in buildings has been reported by Abohela et al, 2011. The authors stated that wind turbine integration in buildings should be studied by multi-disciplinary teams in the early stages of building development. They also mentioned that wind generator use should include apart from economic, technical and environmental criteria their social acceptance. A new building-integrated wind turbine system utilizing the building skin has been proposed by Park et al, 2015. The authors have designed a wind generator which utilized increased wind speeds due to specific configuration of building skin to generate electricity achieving a high power coefficient at 0.381. An introduction to small wind turbine project has been presented by Forsyth, 1997. The author described the efforts made to improve small wind turbines (5-40 KW) in order to have high reliability and low maintenance. He stated that the cost of electricity generated was at 0.6 $/KWh for average wind speeds at 5.5 m/s.

2.6 Solid biomass use

An economic assessment of biomass boiler plants in Italy has been presented by Carlini et al, 2013. The authors have assessed the economic viability of using biomass boiler plants in residential buildings in Viterbo, Italy. They stated that governmental subsidy of biomass heating systems has resulted in low payback periods for these green energy investments in residential buildings. A review of biomass heating in UK homes and various commercial applications has been reported by Dwyer et al, 2006. The authors suggested that heating of buildings with biomass should be studied in the early stages of building design to achieve maximum benefits. However, they mentioned, biomass heating should not be considered as a standard solution like oil or gas for heating buildings but rather as a green initiative to them.

2.7 High efficiency heat pumps

A general review of ground-source heat pump (GSHP) systems for space heating and cooling in buildings has been presented by Sarbu et al, 2014. The authors stated that GSHPs are suitable for heating and cooling in buildings having high efficiencies and reduced CO_2 emissions. The annual increase in their use in buildings is remarkable due to the fact that they are more efficient than many traditional energy systems. An evaluation of GSHP systems for residential buildings in warm Mediterranean regions like Cyprus has been reported by Michopoulos et al, 2016. The authors have analyzed with appropriate software two buildings, a single- and a multi-family building for two cases. In the former building a GSHP was used while the latter used a traditional system. Their results indicated a lower primary energy use in both buildings and a lower cost in the multi-family building when the GSHP was used. CO_2 emissions were higher when the GSHP was used in the single-house building but lower in the multi-family building. A study on energy performance of a GSHP operation in historical buildings in Italy has been reported by Pacchiega et al, 2017. The authors stated that refurbishment of historical buildings, due to their characteristics, presents various difficulties regarding the selection of the optimal energy system. They also mentioned that GSHPs consist of the best solution, for heating and cooling, among those analyzed in historical buildings.

2.8 Co-generation of heat and power

Martinez et al, 2011 have reported on energy supply in buildings with micro-cogeneration and heat pumps. The authors stated that these energy efficient technologies are important regarding energy supply in buildings for achieving the goals of EU legislation. An evaluation of combined heat and power generation has been published by International Energy Agency, 2008. According to this report penetration of CHP technologies in buildings is rather limited due to various barriers. These include electricity grid generation interconnection regulations, lack of knowledge about CHP benefits and savings and lack of integrated urban heating and cooling supply planning.

2.9 District heating using biomass

The design of a biomass district heating system has been reported by Vallios et al, 2007. The authors presented a detailed design methodology of such energy systems which offer the possibility for their technical and economic assessment since there is limited experience and only few applications in Southern Europe. A review on biomass use in Swedish district heating systems has been presented by Ericsson et al, 2016. The authors stated that district heating satisfies about 60% of the heat demand in Swedish buildings while biomass alone accounts for about half of their heat supply. They concluded that biomass introduction and expansion was supported by national energy policy tools and local municipal initiatives. The use of biomass in district heating has been studied by Chatzistougianni et al, 2016. The authors stated that locally available biomass can support a small-scale district heating system of public buildings especially when taking into account energy audits, in-situ measurements and energy efficiency improvement measures.

2.10 District heating using waste heat

District heating using waste heat from thermal power plants in EU has been studied by Colmenar-Santos et al, 2016. The authors stated that thermal power plants in EU reject a large amount of heat compared with heat energy used in buildings. They also mentioned that about half of the installed capacity in EU's conventional thermal power plants is located at an appropriate distance from cities and towns allowing the use of the waste heat in district heating systems. Low temperature district heating for future energy systems has been reported by Schmidt et al, 2017. The authors stated that low temperature district heating can significantly contribute to a more efficient use of energy resources as well as better integration of renewable energies, like geothermal energy, solar energy and surplus heat, like industrial waste heat, in the building sector.

2.11 Creation of net zero energy buildings with net zero carbon emissions due to energy use

A study on zero energy balance buildings in southern Italy has been presented by Ferrante et al, 2011. The authors stated that zero energy balance buildings can be achieved combining solar passive systems with solar and wind energy microgeneration. They concluded that zero energy balance and zero on-site CO_2 emissions houses in Mediterranean climate are easily accessible goals. A study of net zero CO_2 emissions residential buildings, due to energy use, located in Crete, Greece has been reported by Vourdoubas, 2016. The author indicated that combined use of solar thermal energy, solar-PV and solid biomass could cover all the energy needs in a building zeroing its CO_2 emissions due to energy use. He also stated that the same result is achieved in the building with combined use of solar thermal energy, solar-PV and GSHPs.

The aims of the current work are:
a)   To review the sustainable energy technologies which are currently used in buildings for providing heat, cooling and electricity in Mediterranean basin,
b)   To assess them with technical and non-technical criteria, and
c) To investigate if their combined use could result in the creation of a net zero energy building with net zero carbon emissions due to operating energy use.

3. Use of sustainable energy technologies for energy generation in buildings

Various sustainable energy technologies including renewable energies and efficient energy systems can be used for providing energy in buildings in Mediterranean region. They can generate heat, cooling and electricity and their use depends on many parameters including their availability, local climate conditions and their cost-effectiveness. Some of them are used broadly while others have limited applications. New innovative green energy systems are also used experimentally and they will probably be expanded and commercialized in the future. Solar energy is used in buildings for heat, cooling and electricity generation. Solar thermal systems with flat plate collectors have been used for many decades while co-generation systems with parabolic collectors have been recently developed without proving their cost-effectiveness so far. Solar-PV systems are broadly used today after a sharp decrease in their prices in previous years. Solar cooling systems are only used in large buildings although this technology is very challenging in Mediterranean region. Small wind turbines have limited applications mainly in off-grid buildings. Solid biomass burning is broadly used for heat generation particularly in rural areas. Heat pumps are increasingly used for heating and cooling in buildings having many advantages including their high energy efficiencies. Co-generation of heat and power fuelled by natural gas is mainly used in large buildings. Finally district heating systems fuelled with either biomass or waste heat are used only when fuels or energy are available nearby the building's area. A list of various sustainable energy technologies which could be used for energy generation in buildings is presented in Table 1.

Table 1. Various sustainable energy technologies which can be used in buildings for generation of heat, cooling and electricity in Mediterranean basin[1]

	Technology	Space heating	Space cooling	Domestic hot water production	Electricity
1	Solar thermal with flat plate collectors			+	
2	Solar thermal with parabolic collectors and Sterling engine	+		+	+
3	Solar thermal cooling		+		
4	Solar-PV				+
5	Small wind turbines				+
6	Solid biomass burning	+		+	
7	High efficiency heat pumps	+	+	+	
8	Co-generation of heat and power	+		+	+
9	District heating	+		+	

	using solid biomass				
10	District heating using waste heat	+		+	
	Total	6	2	7	4

¹*Source: Own estimations*

Table 1 indicates that many green energy technologies can provide heat but only few can provide cooling. Current use of sustainable energy technologies in buildings in Mediterranean region is presented in Table 2.

Table 2. Current use of various sustainable energy technologies in buildings in Mediterranean area [1]

	Technology	Current use of the technology
1.	Solar thermal with flat plate collectors	Yes, broadly
2.	Solar thermal with parabolic collectors and Sterling engine	Experimentally, few commercial applications
3.	Solar thermal cooling	Limited applications mainly in large buildings
4.	Solar-PV	Yes, broadly
5.	Small wind turbines	Limited applications mainly in off-grid buildings
6.	Solid biomass burning	Yes, broadly
7.	High efficiency heat pumps	Yes, broadly
8.	Co-generation of heat and power	Limited applications mainly in large buildings
9.	District heating with solid biomass	Limited applications
10.	District heating with waste heat	Limited applications

¹*Source: Own estimations*

The efficiency of various sustainable energy systems varies significantly since some technologies like solar-PVs have low efficiency while others like GSHPs and co-generation systems have higher efficiencies. The efficiency of various technologies used in buildings is presented in Table 3.

Table 3. Efficiency of various sustainable energy technologies used in buildings in Mediterranean area[1]

	Technology	Efficiency (%)
1	Solar thermal with flat plate collectors	30-35
2	Solar thermal with parabolic collectors and Sterling engine	There are not enough reliable data
3	Solar thermal cooling	100-120
4	Solar-PV	15-18
5	Small wind turbines	Up to 40%
6	Solid biomass burning	70-80
7	High efficiency heat pumps	250-350 (COP =2.5-3.5)
8	Co-generation of heat and power	75-85
9	District heating with solid biomass	70-80
10	District heating with waste heat	80-90

[1]Source: Own estimations

4. Assessment of various sustainable energy technologies use in buildings

Assessment of the abovementioned renewable energy technologies has been made using technical and non-technical criteria. Among the non-technical criteria are included the simplicity of the installation of the energy system, its accessibility, its easiness of use from the end user and its social acceptance. Among the technical and economic criteria are included its cost-effectiveness, its easiness in maintenance, its reliability and its efficiency. For each criterion a score between 1 (low) to 5 (high) is given. Scores for non-technical assessment and for technical and economic assessment are presented in Tables 4 and 5.

Table 4. Non-technical assessment of various sustainable energy technologies used in buildings in Mediterranean area [1,2]

	Technology	Simplicity of installation	Accessibility	Easiness to use for the end user	Social acceptance	Total score
1	Solar thermal with flat plate collectors	5	4	5	5	19

2	Solar thermal with parabolic collectors and Sterling engine	1	2	2	4	9
3	Solar thermal cooling	2	2	4	4	12
4	Solar-PV	5	4	5	5	19
5	Small wind turbines	2	3	3	2	10
6	Solid biomass burning	4	4	4	4	16
7	High efficiency heat pumps	5	5	5	5	20
8	Co-generation of heat and power	4	3	4	4	15
9	District heating using solid biomass	5	5	5	5	20
10	District heating using waste heat	5	5	5	5	20

[1]Score: 1=low to 5=high, [2]Source: Own estimations

Table 5. Technical and economic assessment of various sustainable energy technologies used in buildings in Mediterranean area [1,2]

	Technology	Cost effectiveness	Easiness to maintain	Reliability	Efficiency	Total score
1.	Solar thermal with flat plate collectors	4	5	5	4	18
2.	Solar thermal with parabolic collectors and	2	1	3	-	6

	Sterling engine					
3.	Solar thermal cooling	2	2	3	5	12
4.	Solar-PV	4	5	5	2	16
5.	Small wind turbines	2	2	4	2	10
6.	Solid biomass burning	5	4	5	5	19
7.	High efficiency heat pumps	4	4	5	5	18
8.	Co-generation of heat and power	4	4	5	5	18
9.	District heating using solid biomass	5	5	5	5	20
10.	District heating using waste heat	5	5	5	5	20

[1]Score: 1=low to 5=high, [2]Source: Own estimations

The total score and the ranking of sustainable energy technologies are presented in Table 6. The best scores are achieved in district heating systems using either solid biomass or waste heat since the district heating operator is providing heat into the buildings which are the end users while the heat price is relatively low.

Table 6. Ranking of the energy technologies according to the score given[1]

	Technology	Score for the non-technical criteria	Score for the technical and economic criteria	Total score	Ranking
1.	Solar thermal with flat plate collectors	19	18	37	3rd
2.	Solar thermal with parabolic collectors and Sterling engine	9	6	15	8th

3.	Solar thermal cooling	12	12	24	6th
4.	Solar-PV	19	16	35	4th
5.	Small wind turbines	10	10	20	7th
6.	Solid biomass burning	16	19	35	4th
7.	High efficiency heat pumps	20	18	38	2nd
8.	Co-generation of heat and power	15	18	33	5th
9.	District heating using solid biomass	20	20	40	1st
10.	District heating using waste heat	20	20	40	1st

5. Creation of net zero energy buildings in Mediterranean basin

Creation of buildings with net zero energy balance and net zero CO_2 emissions due to energy use consists of a great challenge nowadays for mitigating climate change. Reduction of energy consumption in buildings using various energy saving techniques combined with in-situ energy generation from benign green energy sources could achieve the abovementioned goal. Various studies implemented so far indicated its feasibility due to technology improvements and to cost reduction of sustainable energy technologies. Various renewable energies which, combined, can provide all the energy required in a grid-connected building are presented in Table 7. Solar thermal energy can provide domestic hot water while solar-PV energy the electricity required for lighting and operation of various electric equipment according to net-metering regulations. Solid biomass can provide space heating while GSHPs can provide space heating and cooling as well as DHW production. Apart from the renewable energies presented in table 7 combined use of different renewable energy technologies could achieve the same goal. District heating systems fuelled with solid biomass or waste heat can provide space heating and domestic hot water, heat pumps can provide space cooling and solar-PV panels can provide all the electricity required in the building. All these technologies are mature, reliable, well-proven and cost-effective. Renewable energies mentioned are abundant in Mediterranean region except the limited availability of solid biomass in southern Mediterranean countries.

Table 7. Renewable energy technologies which can provide all the energy required in a building [1,2]

Renewable energy	Combined use of solar thermal, solar - PV energy and solid biomass	Combined use of solar thermal, solar-PV energy and GSHPs
Solar thermal	Domestic hot water	Domestic hot water
Solar-PV	Electricity generation	Electricity generation
Solid biomass	Space heating, domestic hot water	
Low enthalpy geothermal energy		Space heating and cooling, domestic hot water

[1]Grid connected building with net-metering regulations, [2]Source: Vourdoubas, 2016

6. Discussion

Renewable energy technologies which can be used for heat, cooling and electricity generation like solar energy, wind energy, low enthalpy geothermal energy and solid biomass are abundant in many Mediterranean countries. Various mature and reliable renewable energy technologies can be used without requiring financial subsidies. Current EU regulations make obligatory the construction of nearly zero energy buildings while policies like net-metering regulations promote the use of solar-PVs or co-generation systems in buildings without direct economic subsidies. Biomass-fuelled district heating systems have very good performance but they currently operate only in northern Mediterranean countries in areas with high availability of biomass resources. Solar thermal systems with parabolic collectors and Sterling engines have only recently penetrated into the market while evaluation of their performance needs more time. Energy renovation of existing buildings in order to reduce their energy consumption and their carbon footprint due to energy use contributes in mitigation of climate change while decreases the dependence of various countries on imported fossil fuels. Solar energy is abundant in Mediterranean region and the use of solar-PVs is currently promoted in many countries. Only two technologies are available for space cooling in buildings. One only though is being used extensively today. Probably technological improvements in the near future will allow the creation of more reliable and cost-effective solar cooling systems which could be used in Mediterranean countries which have abundant solar energy resources. Higher use of sustainable energy technologies in buildings does not require financial subsidies but rather non-financial support and removal of various barriers. These include appropriate training of engineers and building constructors concerning the use of renewable energies in buildings, involvement of the private sector and third party financing in energy investments particularly in large public buildings and/or in social houses, better information for consumers regarding the benefits of sustainable energies use as well as creation of few pilot public buildings with net zero energy consumption which could raise awareness among citizens and they will demonstrate the use of benign green energy sources in them.

7. Conclusions

Various mature, reliable and cost-effective renewable energies can be used for heat, cooling and power generation in buildings in Mediterranean region. These include solar thermal energy with flat plate collectors, solar thermal energy with parabolic collectors, solar cooling, solar-PV energy, solid biomass burning, small wind turbines, high efficiency heat pumps, co-generation of heat and power and district heating systems using either solid biomass or waste heat. Solar energy, low enthalpy geothermal heat and, at least in some countries, solid biomass are abundant in this geographical district. The majority of the above-mentioned renewable energy technologies are related with heat generation in buildings. Current policies and the legal framework in most Mediterranean countries support their use for energy generation in them. The efficiency of the sustainable energy technologies examined varies broadly between 15 % and 350 %. Various criteria have been used for evaluation of sustainable energy technologies. These include non-technical criteria like accessibility, social acceptance, simplicity in installation and easiness to use by the end user. Additionally, technical and economic criteria include reliability, efficiency, easiness in maintenance and cost-effectiveness. Assessment of examined technologies used for energy generation in buildings in Mediterranean area indicated that district heating systems, solar thermal systems with flat plate collectors, high efficiency heat pumps and solar-PV panels have the best performance. Combined use of various renewable energy technologies in Mediterranean buildings could zero their net energy use and their net carbon emissions. These technologies are mature, reliable and cost-effective while they are already used separately in various buildings. Therefore, creation of net zero energy buildings is currently feasible and cost-effective in Mediterranean basin. Further work should be focused on the transformation of various public buildings in Mediterranean countries to net zero energy buildings with the combined use of the abovementioned renewable energy technologies.

References

[1] Abohela, I., Hamza, N. & Dudek, S. (2011). "Urban wind turbines integration in the build form and environment", *in the Forum Ejournal 10*, June 2011, pp. 23-39. Retrieved at 4/6/2918 from https://research.ncl.ac.uk/forum/v10i1/2_Islam.pdf

[2] Aelenei, L., Smyth, M., Platzer, W., Norton, B., Kennedy, D., Kalogirou, S. & Maurer, Ch. (2016). "Solar thermal systems - Towards a systematic characterization of building integration", *Energy Procedia*, 91, pp. 897-906. doi: 10.1016/j.egypro.2016.06.256

[3] Baldwin, Ch. & Cruickshank, C.A. (2012). "A review of solar cooling technologies for residential applications in Canada", *Energy Procedia*, 30, pp. 495-594. doi: 10.1016/j.egypro.2012.11.059

[4] Biyik, E., Araz, M., Hepbasli, A., Shahrestani, M., Yao, R., Shao, L., Essah, E., Oliveira, C.A., Del Cano, T., Rico.E., Lechon, J.L., Andrade, L., Mendes, A. & Atli, Y.B. (2017). "A key review of building integrated photovoltaic (BIPV) systems", *Engineering Science and Technology, An International Journal,* 20, pp. 833-858. https://doi.org/10.1016/j.jestch.2017.01.009

[5] Bobrova, D. (2015). "Building-Integrated wind turbines in the aspect of Architectural shaping", *Procedia Engineering*, 117, pp. 404-410. doi: 10.1016/j.proeng.2015.08.185

[6] Carlini, M., Castellucci, S., Cocchi, S., Allegrini, E. & Li, M. (2013). "Italian residential buildings: Economic assessment for biomass boilers plants", *Mathematical Problems in Engineering,* Article 823851, http://dx.doi.org/10.1155/2013/823851

[7] Chatzistougianni, N., Giagozoglou, E., Sentzas, K., katastergios, E., Tsiamitros, D., Stimoniaris, D., Stomoniaris, A. & Maropoulos, S. (2016). "Biomass district heating methodology and pilot installations for public buildings groups", in *20th Innovative Manufacturing Engineering and Energy Conference*. doi:10.1088/1757-899X/161/1/012083

[8] Colmenar-Santos, A., Rosales-Asensio, E., Borge-Diez, D. & Blanes-Peiro, J.J. (2016). "District heating and co-generation in the EU-28: Current situation, potential and proposed energy strategy for its generalization", *Renewable and Sustainable Energy Reviews*, 62, pp. 621-639. http://dx.doi.org/10.1016/j.rser.2016.05.004

[9] Dwyer, T. (2006). "A review of biomass heating for UK homes and commercial applications", *International Journal of Low Carbon Technologies*, 1(4), pp. 329-335. https://doi.org/10.1093/ijlct/1.4.329

[10] Ericsson, K. & Werner, S. (2016). "The introduction and expansion of biomass use in Swedish district heating systems", *Biomass and Bioenergy*, 94, pp. 57-65. http://dx.doi.org/10.1016/j.biombioe.2016.08.011

[11] Ferrante, A. & Cascella, M.T. (2011). "Zero energy balance and zero on-site CO_2 emission housing development in Mediterranean climate", *Energy and Buildings*, 43, pp. 2002-2010. doi:10.1016/j.enbuild.2011.04.008

[12] Forsyth, T.L. (1997). "An introduction to the small wind turbine project", presented at *Wind-power '97*, Austin, Texas, USA, June 15-18, 1997, National renewable energy laboratory.

[13] Greening, B. & Azapagic, A. (2014). "Domestic solar thermal water heating: A sustainable option for the UK?", *Renewable Energy*, 63, pp. 23-36. http://dx.doi.org/10.1016/j.renene.2013.07.048

[14] Grignaffini, S. & Romagna, M. (2012). "Solar cooling: a case study", *WIT transactions on ecology and the environment, Eco-Architecture IV*, 165, pp. 399-410. doi:10.2495/ARC120351

[15] Hafez, A.Z., Soliman, A., El-Metwally, K.A. & Ismail, I.M. (2016). "Solar parabolic disc Sterling engine system design, simulation and thermal analysis", *Energy Conversion and Management*, 126, pp. 60-75. DOI: 10.1016/j.enconman.2016.07.067

[16] Hayter, S., Torcellini, P. & Deru, M. (2002). Photovoltaics for buildings: New applications and lessons learnt, American council for an energy efficient economy, *Summer study on energy efficiency in buildings*, Pacific Grove, California, USA, August 18-23, 2002. Retrieved at 4/6/2018 from https://www.nrel.gov/docs/fy02osti/32158.pdf

[17] Henning, H.M. & Doll, J. (2012). "Solar systems for heating and cooling of buildings", *Energy Procedia*, 30, pp. 633-653. doi: 10.1016/j.egypro.2012.11.073

[18] International Energy Agency, "Combined Heat and Power, Evaluating the benefits of greater global investments", France, 2008. Retrieved at 4/6/2018

from https://www.iea.org/publications/freepublications/publication/chp_report.pdf

[19] International Energy Agency, Task 38, "Solar air conditioning and refrigeration", Solar Cooling position paper, 2011. Retrieved at 4/6/2018 from https://nachhaltigwirtschaften.at/resources/iea_pdf/iea_shc_solar_cooling_position_paper.pdf

[20] Martinez, M.G., Andrade, L.C., Bugallo, P.M.B. & Iglesias, M.B. (2011). "Energy supply in buildings: heat pump and micro-cogeneration", In *World Renewable Energy Congress 2011*, 8-13 May 2011, Linkoping, Sweden.

[21] Michopoulos, A., Voulgari, V., Tsikaloudaki, A. & Zachariadis, Th. (2016). "Evaluation of ground source heat pump systems for residential buildings in warm Mediterranean regions: the example of Cyprus", *Energy Efficiency,* 9, pp. 1421-1436. DOI 10.1007/s12053-016-9431-1

[22] Pacchiega, C. & Fausti, P. (2017). "A study on the energy performance of a ground source heat pump utilized in the refurbishment of an historical building: comparison of different designed options", *Energy Procedia*, 133, pp. 349-357. https://doi.org/10.1016/j.egypro.2017.09.360

[23] Park, J., Jung, H.J., Lee, S.W. & Park, J. (2015). "A new building integrated-wind turbine system utilizing the buildings, *Energies*", 8, pp. 11846-11870. doi:10.3390/en81011846

[24] Quintal, E.S., Bernardo, H.S., Amaral, P.G. & Neves, L.D. (2011). "Use of parabolic trough solar collectors for building air conditioning and domestic hot water production - A case study", *3RD International Youth Conference on Energetics,* July 2011. DOI: 10.13140/RG.2.1.2498.7683

[25] Salem, T. & Kinab, E. (2015). "Analysis of building-integrated photovoltaic systems: A case study of commercial buildings under Mediterranean climate", *Procedia Engineering,* 118, 5380545. doi: 10.1016/j.proeng.2015.08.473

[26] Sarbu, I. & Sebarchievici, C. (2014). "General review of ground source heat pump systems for heating and cooling buildings", *Energy and Buildings*, 70, pp. 441-454. http://dx.doi.org/10.1016/j.enbuild.2013.11.068

[27] Schmidt, D., Kallert, A., Blesl, M., Svendsen, S., Li, H., Nord, N. & Sipila, K. (2017). "Low temperature district heating for future energy systems", *Energy Procedia*, 116, pp. 2638. https://doi.org/10.1016/j.egypro.2017.05.052

[28] Valios, I. & Tsoutsos, Th. (2007). "Design of biomass district heating systems", In *15th European Biomass Conference*, Berlin, Germany, 7-11 May 2007.

[29] Vinubhai, T.S., Vishal, R.J. & Thakkar, K. (2014). "A review: Solar water heating systems", in the *National conference on emerging Vista of technology in the 21st century*. doi: 10.13140/2.1.1910.5281

[30] Vourdoubas, J. (2016). "Creation of zero CO_2 emission residential building due to energy use: A case study in Crete, Greece", *Journal of Civil Engineering and Architecture Research*, 3(2), pp. 1251-1259.

[6] Creation of net zero carbon emissions residential buildings due to energy use in Mediterranean region. Are they feasible?

1. Introduction

Buildings consume large amounts of energy and emit GHGs into the atmosphere. The necessity to mitigate climate change requires the sharp decrease in fossil fuels consumption and the resulting GHGs emissions. New EU regulations promote nearly-zero energy buildings (NZEBs) making obligatory the low energy consumption in new buildings as well as the increase of energy efficiency in old buildings which should be energy renovated. Buildings consume energy during their construction, operation, refurbishment and demolition. Building's embodied energy includes the energy used during their construction, refurbishment and demolition. When buildings' operating energy is reduced the share of embodied energy in their total life cycle energy use is increased. Apart from the building sector the challenge to reduce fossil fuel consumption in the transportation sector necessitates replacement of conventional vehicles, having internal combustion engines, with electric vehicles (EVs). The electric batteries of these vehicles can be recharged with solar electricity generated with solar photovoltaic (solar-PV) systems installed in residential buildings. Solar energy is abundant in Mediterranean region and it can be used for heat and electricity generation with solar thermal and solar-PV systems. The required solar energy technologies are mature, reliable and cost-effective while they are already used commercially in buildings as well as in other applications. Promotion of net zero carbon emissions green buildings (NZCEBs) in Mediterranean basin using locally available benign energy sources for covering their embodied and operating energy needs as well as the electricity required in residents' electric cars could assist in achievement of global targets regarding carbon emissions reduction. Various existing barriers should be removed for future promotion of NZCEBs. The necessary sustainable energy technologies are mature, well proven, reliable and cost-effective while the necessary legislative framework is favorable in many countries.

2. Literature survey

2.1 Energy consumption in buildings

Calculation of energy consumption and carbon emissions in housing construction in Japan has been presented [1]. The authors stated that embodied energy in residential buildings depends on the type of construction and the materials used.

They estimated that their embodied energy varied between 833-8,777 KWh/m^2 while their carbon emissions due to embodied energy was at 250-850 kgCO$_2$/m^2. Energy consumption in EU and Hellenic buildings has been reported [2]. The authors mentioned that total annual energy consumption in EU buildings varied between 150-230 KWh/m^2. They also stated that annual residential energy use per capita varied from 1,500-5,000 KWh/capita in Southern Europe up to 8,000 KWh/capita in Northern Europe. A study on the energy performance of existing dwellings has been published [3]. The authors mentioned that in EU the final energy consumption of the building sector corresponds at 40.3 % of the total EU-25 final energy use while the energy consumed in dwellings corresponds at 25.4% of the total final energy use. A study on energy consumption in buildings has been published [4]. The authors stated that energy consumption varies significantly according to end uses among EU countries while the largest share of energy consumption corresponds to air-conditioning.

2.2 Embodied and operating energy consumption

A report regarding carbon emissions due to embodied energy of buildings has been released [5]. It stated that the impacts of embodied energy in buildings become greater with decrease of their operating energy while increasing their thermal insulation has minimal impacts on their embodied energy related carbon emissions. It also mentioned that the concrete structure has the highest amount of embodied carbon followed by steel and timber. An analysis on life cycle energy consumption in buildings has been published [6]. The authors mentioned that life cycle primary energy requirements of conventional residential buildings falls in the range of 150-400 KWh/m^{2}year which includes both operating and embodied energy. Operating energy corresponds at 80-90% of the total energy use while the embodied energy at 10-20%. They also reported that demolition energy has a negligible effect on the total energy balance of the building. A report on the operating and embodied energy of an Italian building has been published [7]. The authors emphasized the key issue of embodied energy which is particularly important in the case of low energy buildings. They also pinpointed the difficulty in defining the reference area in the building including service and unheated zones as well as the absence of an internationally accepted protocol for that. A life cycle energy analysis in buildings has been implemented [8]. With reference to net zero energy buildings, where on-site renewable energy generation covers the annual energy load, the authors mentioned the increase of the share of embodied energy compared with operating energy. They stated that: a) in the last decades the embodied energy in new buildings has slightly decreased, and b) the relative share of embodied energy to life cycle energy use has significantly increased due to the sharp decrease in operating energy consumption because of more strict energy regulations. An investigation on the possibility of creating NZCEBs in Crete, Greece due to life cycle energy use has been made [9]. The author stated that creation of net zero CO_2 emissions residential buildings due to life cycle energy use in Crete, Greece does not have major difficulties and it could be achieved relatively easily with the use of mature, reliable and cost effective renewable energy technologies. A review of current trends regarding operating versus embodied energy related carbon emissions in buildings has been published [10]. The authors stated that in order to mitigate climate change buildings must be designed and constructed with minimum environmental impacts. Total life cycle carbon emissions from buildings are both due to operating and embodied energy use. Considerable efforts have been made to reduce operating energy related carbon emissions but little attention has been paid to embodied energy related carbon emissions. Therefore, a critical review of the relation between operating and embodied emissions is necessary in order to highlight the importance of embodied energy related carbon emissions. A report on life cycle energy consumption in net zero energy buildings has been released [11]. The authors stated that apart from the operating energy the energy used during its construction, refurbishment and demolition should be taken into account.

2.3 Net zero energy buildings

An analysis of a conventional house and a net zero energy house has been performed [12]. The authors proposed the use of energy saving systems, a solar water heater and solar-PV panels in net zero energy houses. They mentioned that many of these sustainable energy systems are cost-effective. A presentation of myths and facts regarding zero energy and zero carbon emissions buildings has been made [13]. The author stated that in recent years these types of buildings have attracted much attention in many countries although there is a lot of debate as to whether their construction is feasible. He concluded that energy self-sufficiency in buildings can be achieved and net zero energy and net zero carbon emissions buildings will soon become economically and socially accepted. A report on net-positive energy buildings has been released [14]. The author stated that net-positive energy buildings introduced several new design considerations and possibilities. He mentioned that net-positive energy buildings involve energy and economic exchanges and negotiations with power utilities. A review of the research already published regarding low or zero carbon emissions buildings has been published [15]. The authors have considered the use of renewable energy technologies, use of low carbon building materials as well as the design and assessment of methods used. Among various renewable energy technologies they studied wind turbines, solar-PVs, solar thermal collectors, wood heating and high efficiency heat pumps. A holistic study on the concept of zero energy buildings has been realized [16]. The author mentioned that firstly energy saving measures should be taken and secondly renewable energies should be utilized in order to supply the required energy in buildings. It was also stated that it is always easier and environmentally friendly to save energy than to produce it. It was also suggested that indoor climate conditions should be defined in order to compare zero energy buildings (ZEBs) in different locations. A critical look at zero energy buildings has been made [17]. The authors have defined four types of ZEBs as follows: (a) Net zero site energy; (b) Net zero source energy; (c) Net zero energy cost; and (d) Net zero energy emissions. They have also commented on the advantages and disadvantages of each one of them. A review on net zero energy buildings has been presented [18]. The authors stated that net zero energy buildings are the future target in building's design and construction while this requires a clear and consistent definition as well as a commonly agreed methodology for the calculations.

2.4 Net zero carbon emissions buildings

A report on feasibility of zero carbon emission homes in England in 2016 from the perspective of house builders has been published [19]. The authors stated that achieving the targets for carbon emissions in UK by 2050 all industries including the housing sector must reduce their carbon emissions. They mentioned that although net zero carbon homes are technically feasible in the long term, clear and concise actions are required from both the government and the house building industry. A study on the existing barriers for constructing zero carbon emission homes in UK has been realized [20]. The authors mentioned that, from the point of view of the construction industry, five barriers have been identified which are categorized as economic, skills and knowledge, industry, legislative and cultural. They stated that although the barriers are more than the drivers in the zero carbon emissions homebuilding industry new policy mechanisms could overcome them. A study on sustainable design for zero carbon architecture has been made [21]. The authors mentioned that zero carbon homes are expected to decrease their energy requirements via effective "passive and active design solutions" and secondly by means of renewable energy systems to supply the remaining energy demand. They also stated that focus should be on "building's envelope". A study on the cost of carbon emissions reduction in new buildings has been reported [22]. The study indicated that an additional capital cost at 5-11% of the initial building's construction cost is required in order to achieve the zero carbon emissions targets. A report concerning net zero carbon emissions buildings has been published [23]. The report mentioned that embodied energy in new buildings has a share of approximately 50% in their life cycle energy consumption. It is also stated that five steps should be followed in achieving a net zero carbon emissions building which includes planning, reduction of construction impacts, reduction of operational impacts, increase in renewable energy supply and offset of any remaining carbon emissions. Creation of net zero CO_2 emissions residential buildings due to operating energy use in Crete, Greece has been reported [24]. The author stated that using reliable and cost-effective renewable energy technologies including solar thermal, solar-PV, solid biomass combustion and ground-source heat pumps could cover all its operating energy requirements. A study regarding zero carbon building refurbishment has been realized [25]. The authors categorized a range of energy technologies in a hierarchical manner for building's energy renovation. The proposed hierarchical pathway of sustainable energy technologies included building insulation, high efficiency equipments and micro-generation using renewable energy technologies.

2.5 Use of solar electricity to charge batteries in electric cars

An investigation of the possibilities of recharging electric car's batteries with solar-PV systems installed in residential buildings in Sweden has been implemented [26]. The authors mentioned that home charging of electric batteries increases self-consumption of solar electricity. They stated though that due to climate conditions this option is not attractive in Sweden while it would be an interesting solution in countries with high solar irradiance throughout the year. Research on the possibility of recharging electric vehicle's batteries with solar energy in workplaces in Netherlands has been implemented [27]. The authors mentioned that due to low solar irradiance in the country solar-PV panels should be oversized with respect to converter's power. They also stated that the solar-PV charger can integrate a storage system in order to be independent from the grid. A report on smart EV charging systems in Norway has been published

[28]. The report investigated the interaction of charging stations with energy requirements in buildings and local electricity generation. Solar-PV systems installed in homes can be used for car batteries recharging since they increase electricity self-consumption. Power use during batteries recharging varies between 2.3 KW to 3.6 KW while fast recharging requires higher power.

The aims of the current study are:

a)   The investigation of the possibility of creating NZCEs residential buildings in Mediterranean region with reference to their requirements in operating and embodied energy as well as the energy consumed for recharging the batteries of residents' electric cars,

b)   The presentation of the appropriate sustainable energy technologies which could be used in these buildings, and

c)   The cost estimation of the required sustainable energy systems for achieving this goal.

The methodology followed includes: 1. Estimation of energy consumption in a residential building including its operating energy, embodied energy and energy used in recharging electric batteries of resident's vehicles, 2. Presentation of the characteristics of the reliable, mature and cost-effective renewable energy technologies which could be used, 3. Presentation of a case study for a NZCE residential building and sizing of the necessary sustainable energy systems, and 4. Cost and environmental considerations.

3. Energy requirements in residential buildings

Energy consumed in a residential building over its life span includes the energy used in construction, operation, refurbishment and demolition. Energy consumed during the phases of construction, refurbishment and demolition is defined as the embodied energy of the building. Operating energy in a residential building is the energy consumed during its operation. Embodied energy has a low share in its life cycle energy consumption while operating energy has the highest share in the total energy use over the life span of the building.

3.1 Operating energy use

Energy is consumed in various sectors of residential buildings including:
1.   Space heating,
2.   Space cooling,
3.   Domestic hot water (DHW) production,
4.   Lighting, and

5. Operation of various electric appliances and apparatus

The main energy sources used are grid electricity for lighting, operation of electric devices and air-conditioning, while fossil fuels, mainly diesel oil and natural gas, are often used for space heating and DHW production. The typical operating energy consumption in a residential building with a covered area at 120 m² located in the island of Crete, Greece is presented in Table 1.

Table 1. Typical operating energy consumption in a residential building located in the island of Crete, Greece [1]

Sector	Annual specific energy requirements (KWh/m²)	Annual energy requirements (KWh)	% (Energy requirements)	Annual specific CO₂ emissions (kgCO₂/m²)	Annual CO₂ emissions (kgCO₂)	% (CO₂ emissions)
Space heating [2]	107.1	12,852	63	33.20	3,984	44.85
Space cooling [3,4]	11.9	1,428 [4]	7	2.55	306	3.45
Lighting	20.4	2,448	12	15.30	1,836	20.68
Operation of various electric devices	15.3	1,836	9	11.47	1,377	15.51
DHW production [5]	15.3	1,836	9	11.47	1,377	15.51
Total	170	20,400	100	74	8,880	100

[1] Covered area, 120 m², [2] Use of diesel oil, 0.31 kgCO₂/KWh, [3] Use of heat pump with C.O.P.=3.5, [4] Energy consumption by the heat pump, 408 KWh/year, [5] Use of electricity, 0.75 kgCO₂/KWh

Assuming that the life span of the residential building is 50 years then its overall operating energy consumption over this period is estimated at 1,020,000 KWh or

8,500 KWh/m² while its CO$_2$ emissions are estimated at 444,000 kgCO$_2$ or 3,700 kgCO$_2$/m².

3.2 Embodied energy use

Average embodied energy in a typical residential building varies depending on many factors. It has been estimated, in various published studies, at approximately 5-20% of its life cycle energy consumption. Assuming, in the previously mentioned residential building, that its embodied energy corresponds at 15% of its specific life cycle energy use, it is calculated at 30 KWh/m²y. For the abovementioned residential building and for a life span of 50 years, its overall embodied energy use is estimated at 180,000 KWh or 1,500 KWh/m². Therefore, its overall annual specific life cycle energy consumption, including its embodied and operating energy, is 200 KWh/m² while its life cycle energy consumption is 1,200,000 KWh.

3.3 Energy required for recharging batteries of electric vehicles

Replacement of conventional vehicles having internal combustion engines with EVs using either rechargeable batteries or fuel cells is increasing in many countries for environmental and other reasons. EVs require frequent recharging of their batteries which can be done at home. In this case additional electricity is needed in the residential building for battery recharging. It is assumed that home's residents have two electric vehicles and each vehicle travels 15,000 Km annually while their electricity consumption is 0.2 KWh/Km. In that case the annual electricity requirements in electric vehicles are estimated at 6,000 KWh/year. This corresponds to additional annual specific energy consumption in the residential building at 50 KWh/m².

3.4 Nearly zero energy residential buildings

A NZEB is a building which has significantly reduced its operating energy consumption using various energy saving techniques and technologies resulting in lower heat, cooling and electricity use. The necessity to mitigate climate change has increased the efforts to improve the energy efficiency in buildings lowering their energy consumption, their fossil fuels use and their carbon emissions. New regulations, building codes and legislation in many countries have made obligatory the construction of new buildings with nearly zero energy consumption. Old buildings should also be renovated in order to decrease their energy consumption. With reference to the previously mentioned residential building its energy renovation could decrease its annual specific operating energy consumption from 170 KWh/m² to 50 KWh/m². In this case its annual specific life cycle energy use will be at 80 KWh/m² which is significantly lower than the initial estimated consumption at 200 KWh/m². The energy requirements in the abovementioned residential building regarding its embodied energy, operating energy and energy required for recharging the batteries of two EVs are presented in Table 2.

Table 2. Energy requirements in the residential building regarding its embodied energy, operating energy and energy required for recharging the batteries of two electric vehicles

Annual specific energy requirements	Conventional residential building	Nearly zero energy residential building
Embodied energy (KWh/m²)	30	30
Annual operating energy (KWh/m²)	170	50
Embodied and operating energy (KWh/m²year)	200	80
Energy required annually for recharging electric batteries (KWh/m²)	50	50
Total energy including embodied, operating and energy for recharging electric batteries (KWh/m²year)	250	130
Share of embodied energy to total energy (%)	12	23.08
Share of operating energy to total energy (%)	68	38.46
Share of embodied and operating energy to total energy (%)	80	61.54
Share of energy for recharging electric batteries to total energy (%)	20	38.46

4. Use of renewable energy technologies in residential buildings

Various locally available renewable energy sources have been used for providing heat, cooling and electricity in residential buildings. Their technologies are mature, reliable and cost-effective, while their use in buildings results in net zero carbon emissions due to operating energy use. The most common renewable energies used in the Mediterranean region are solar energy, solid biomass and low enthalpy geothermal energy while the technologies used include the followings:

4.1 Solar thermal energy

It is used for DHW production with flat plate solar collectors, providing hot water at 50-70°C depending on local solar irradiance. These systems are simple in operational and maintenance requirements while they have been used commercially in residential and commercial buildings in the last five decades.

4.2 Solar photovoltaic energy

Solar-PVs are used commercially during the last 10-12 years for electricity generation in on-grid and off-grid residential buildings as well as in other applications. Their use has taken-off due to sharp decrease in their prices during the last two decades. Their annual productivity depends on solar irradiance while their requirements in operation and maintenance are very low. National legislation in many countries encourages and facilitates the use of solar-PVs in buildings as well as in other applications.

4.3 Solid biomass

Locally produced solid biomass can be burnt in appropriate wood stoves or fireplaces generating heat used in space heating and DHW production. It has been used for heat generation for many years and various burning systems currently used are efficient, reliable and cost-effective. Solid biomass is usually a cheap, locally available and renewable fuel which can replace fossil fuels for heat generation in residential buildings located in rural areas. However, its use does not result in net zero carbon emissions, like solar energy, due to energy consumed during its processing and transportation.

4.4 Low enthalpy geothermal energy

High efficiency heat pumps including ground source heat pumps are energy efficient devices used extensively in residential buildings for heat and cooling generation. They utilize the heat stored under the ground while they consume electricity generating heat, domestic hot water and cooling. They are reliable devices having a high initial installation cost but, in the long run, they are cost-effective.

4.5 Other sustainable energy technologies

Various other renewable or low carbon energy technologies, when appropriate, can be used in residential buildings. These include wind turbines generating electricity, heat and power co-generation systems generating heat and electricity, district heating systems providing heat as well as heating systems using rejected industrial heat. Other technologies like solar thermal cooling require further development in order to be commercialized. The most often used renewable energy sources in residential buildings in the Mediterranean region are presented in Table 3.

Table 3. Most commonly used renewable energy technologies in buildings providing heat, cooling and electricity in the Mediterranean region [1]

Energy source	Energy technology used in buildings	Energy generation
Solar energy	Solar thermosiphonic systems with flat plate collectors	Heat for space heating and hot water production
Solar energy	Solar-PV panels	Electricity
Solid biomass	Burning in wood stoves and in fireplaces	Heat providing space heating and domestic hot water
Low enthalpy geothermal energy	High efficiency heat pumps	Heat and cooling, air-conditioning and domestic hot water

[1] Source: Own estimations

5. Net zero carbon emission buildings

Buildings consume fossil fuels and grid electricity for covering their energy requirements and they emit CO_2 into the atmosphere. A net zero carbon emissions (NZCE) building is considered the building which does not emit CO_2 due to its operating energy use. Additionally the building that compensates all its operating energy related with carbon emissions with carbon emissions-free energy generated by renewable energy sources in-situ or off-situ. It can be assumed that a typical residential building uses grid electricity which is generated

by fossil fuels. It also uses fossil fuels including diesel oil or natural gas for heat generation. However, a NZCE residential building should:
a)   Replace all fossil fuels use with renewable energies, and
b)   Offset all the grid electricity used annually with electricity generated by renewable energies like solar-PV electricity, generated in-situ or off-situ. If solar electricity, when generated, is not consumed in the building, it will be injected into the grid. Its embodied energy, additionally to its operating energy, can be also offset with green solar electricity generation.

6. Compensation of grid electricity consumption in residential buildings with green electricity generated in them

Grid electricity consumption can be offset with green electricity generated with renewable energies and injected into the grid. This is allowed in many countries according to net-metering regulations. These regulations allow compensation of the annual grid electricity consumption in the building with solar-PV electricity. Solar-PV panels can be installed on-site or off-site generating electricity which is partly consumed in the building, if needed, while the rest is injected into the grid. Electricity balance is made on an annual or bi-annual basis. If the amount of green electricity generation is higher than the grid electricity consumption the owner does not usually get any financial compensation for the surplus electricity injected into the grid. If the solar-PV system has been seized to generate as much electricity as the building consumes annually then its net electricity consumption is zero and assuming that grid electricity is generated with fossil fuels its net carbon emissions are also zero.

7. A case study of a residential building with net zero carbon emissions due to energy use located in Greece

A case study for a residential building with NZCEs due to energy use is presented. The building is located in Greece which has high solar energy resources. Energy requirements of the residential building are covered with: a) A solar thermal system for DHW production, b) A solar-PV system for electricity generation, and c) A high efficiency ground source heat pump for air-conditioning. The building is connected with the electric grid and it compensates all its annual electricity consumption with solar electricity according to net-metering regulations. In order to calculate the required energy systems in the building, the following assumptions have been made:

1. Its covered area is 120 m² and its annual specific operating energy consumption is 170 KWh/m²,
2. Its embodied energy is equal to 15% of its life cycle energy requirements,
3. The residents have two EVs and they recharge their batteries at home. Each car travels 15,000 Km/year and its energy consumption is 0.2 KWh/Km,
4. A solar thermal system with flat plate collectors is producing two thirds (2/3) of the annually required DHW. The rest is produced with an electric heater. The area of the collectors is 2 m². Its installation cost is 450 €/m² of collector area. The cost of an electric heater producing DHW is 150 €,
5. A solar-PV system is generating the required electricity. The solar-PV system generates 1,500 KWh/KW$_p$ annually while its installation cost is 1,200 €/KW$_p$,
6. A ground source heat pump is covering all its air-conditioning requirements. It operates 1,600 hours/year while its C.O.P. is 3.5. Its installation cost is 2,000 €/KW$_{el}$.

7.1 Solar electricity generation covering all its operating energy needs

1. The annual energy requirements for DHW in the building are 1,836 KWh/year. The solar thermal system produces 1,224 KWh/year while the remaining 612 KWh/year are produced with an electric heater. The solar thermal system has flat plate collectors with an area of 2m² while its installation cost is at 900 €. The power of the electric heater is 3 KW$_{el}$ while its installation cost is at 150 €.
2. The annual energy requirements for air-conditioning (space heating and cooling) are 14,280 KWh/year. Air-conditioning will be provided by a high efficiency heat pump with C.O.P. at 3.5. The electricity consumption by the heat pump is 4,080 KWh/year. Its power is 2.55 KW$_{el}$ while its installation cost is 5,100 €.
3. The total electricity requirements in the residential building include its needs for lighting, operation of various electric devices, requirements for the electric heater and the operation of the heat pump. The total energy is 2,448+1,836+612+4,080=8,976 KWh/year. The nominal power of a solar-PV system providing the required electricity annually is 5.98 KW$_p$ and its installation cost is 7,180.8 €.

The size and installation cost of the required energy systems for covering all the operating energy needs in the residential building are presented in Table 4.

Table 4. Size and installation cost of the required energy systems for covering all the operating energy needs [1]

Energy system	Energy generation	Energy use in the building	Size of the energy system	Installation cost (€)
Solar thermal-flat plate collectors	Heat-hot water production	DHW use	Collector's area 2 m^2	900
Electric heater	Heat-hot water production	DHW use	3 KW$_{el}$	150
Solar-PV	Electricity	Lighting, operation of electric devices, operation of a heat pump, operation of an electric heater	Nominal power 5.98 KW$_p$	7,180.8
Ground-source heat pump	Heating and cooling	Air-conditioning	Electric power 2.55 KW$_{el}$	5,100
Total				13,330.8

[1] Covered area = 120 m^2

7.2 Solar electricity generation covering its embodied energy

Additional solar electricity could be generated and injected into the grid for covering the embodied energy of the residential building. In that case the building is considered as a "negative carbon emissions building" since it generates annually more solar electricity than the grid electricity consumed. The construction and operation of this type of building probably requires negotiations and agreements with the power utility regarding its financial compensation. In that case the size of the solar-PV system should be increased. The embodied energy of the residential building is 30 KWh/m^{2}year (Table 2) or 3,600 KWh/year. The additional size of the solar-PV system for generating the embodied energy annually is 2.4 KW$_p$ and its installation cost is 2,880 €.

7.3 Solar electricity generation for recharging the electric batteries of two vehicles

Additional solar electricity should be generated in order to recharge the electric batteries of residents' cars. The required energy for recharging the batteries of the two EVs is 50 KWh/m²year (Table 2) or 6,000 KWh/year. The additional size of the solar-PV system for generating annually the electricity required for recharging the electric batteries is 4 KW_p and its installation cost is 4,200 €.

8. Economic and Environmental considerations

8.1 Cost estimations

A conventional grid-connected residential building located in Crete usually has a DHW production system and an air-conditioning system. However, a solar-PV system generating all the required electricity in the building should be additionally installed according to net-metering regulations. The size of the necessary solar-PV system should be, depending on the energy requirements that it covers, between 5.98 KW_p and 12.38 KW_p, while its cost varies between 7,180.8 € (59.84 €/m²) and 14,860.8 € (123.84 €/m²).

8.2 Environmental considerations

Use of renewable energy technologies in the residential building will result in reduction of carbon emissions due to energy use. For the calculation of the environmental benefits the following assumptions are made:

1. The building initially uses a solar thermal system for DHW production, electric energy for producing part of the DHW required, lighting and operation of the electric devices including space cooling, while it uses diesel oil for space heating. The C.O.P. of the heat pump used in space cooling is 3.5.
2. CO_2 emissions due to energy use are 0.75 kgCO_2/KWh,
3. CO_2 emissions due to diesel oil use are 0.31 kgCO_2/KWh

Three different scenarios have been considered regarding the use of solar-PV panels for: a) covering only its operating energy demand, b) covering its operating and embodied energy demand, and c) covering its embodied, operating and energy demand for recharging the batteries of two electric cars. The annual CO_2 savings have been calculated at 8,880 kgCO_2 to 16,080 kgCO_2 or 74 kgCO_2/m² to 134 kgCO_2/m². The solar-PV cost per annual carbon emissions savings varies between 0.81 €/kgCO_2 to 0.92 €/kgCO_2. The cost and the environmental benefits due to use of sustainable energy technologies are presented in Table 5.

Table 5. Cost and environmental impacts of the solar-PV system installed in a residential building zeroing its net carbon emissions due to energy use [1]

Solar-PV system	Size of the solar-PV system (KW$_p$)	Cost (€)	Cost (€/m²)	Annual CO$_2$ emissions savings (kg CO$_2$)	Annual specific CO$_2$ emissions savings (kg CO$_2$/m²)
Covering only the operating energy demand	5.98	7,180.8	59.84	8,880	74
Covering the operating and the embodied energy demand	8.38	10,060.8	83.84	11,580	96.5
Covering the operating, embodied and energy demand for recharging the two electric batteries	12.38	14,860.8	123.84	16,080	134

[1] Covered area = 120 m²

Assuming that the construction cost of the abovementioned residential building is at 1,400 €/m² the cost of the required renewable energy systems for its transformation to NZCEs building is estimated at 4.27 % to 8.85 % of its initial construction cost.

9. Discussion

Decreasing energy consumption and carbon emissions in buildings is necessary for achieving the targets for climate change mitigation. Creation of NZCEs buildings require first reduction of their energy consumption and secondly replacement of fossil fuel use with renewable energies. Apart from operating energy use in buildings energy is consumed during their construction, refurbishment and demolition defined as embodied energy. Although in conventional buildings the share of embodied energy is approximately at 15% of their life cycle energy consumption it could reach at 50% in NZEBs. In the present case study the annual energy consumption in the residential building is in the same range of values reported in published literature. Current European policies promote the creation of NZEBs with low energy consumption and low carbon emissions. Various reliable, mature and cost-effective renewable energy technologies which are necessary for creation of NZCEs buildings are already broadly used in various applications. Therefore, their technical feasibility as well as their cost-effectiveness has already been established. Among them solar energy technologies used for heat and electricity generation are very important. Solar energy is abundant in Mediterranean region and it is currently used for energy generation in many applications. The additional cost of the necessary solar-PV system for achieving a NZCEs building has been estimated in previous studies as well as in the current study at 5-10% of the initial construction cost of a typical residential building. Renewable energy technologies necessary for achieving a NZCEs residential building can be used for recharging the batteries of the residents' EVs. In this case self-consumption of solar electricity in the building will be increased. Creation of NZCEs residential buildings will promote energy democracy, increasing the independence of their residents from energy providing utilities. It has been mentioned that energy saving is easier and more desirable than energy generation. However, reduction of energy consumption in the residential building studied has not been considered. If however, the energy consumption is decreased then the size of the required sustainable energy systems would be lower. Various existing barriers hinder the promotion of NZCEs buildings including economic, knowledge and legislative issues. The definition of NZCEs buildings should be clarified while a commonly agreed calculation methodology should be established. Coordination and cooperation of all stakeholders involved in creation of NZCEs buildings including governmental organizations, building designers, engineers, construction companies and the general public, is necessary for their promotion and construction on a large scale.

10. Conclusions

NZCEs residential buildings in Mediterranean region are technically feasible and desirable while the necessary sustainable energy technologies are cost-effective and already commercialized. Renewable energy sources which can be used include solar energy, solid biomass and low enthalpy geothermal energy combined with heat pumps. Solar energy can be used for DHW production and electricity generation, solid biomass for heating and ground source heat pumps for air-conditioning and DHW production. The availability of solar energy in Mediterranean region is high and its use for energy generation is attractive and desirable. The abovementioned renewable energies can cover all the operating and embodied energy requirements in residential buildings as well as the energy required for recharging the batteries of residents' EVs in a cost-effective way zeroing their net carbon emissions. The total installation cost of the required solar-PV system for achieving carbon neutrality due to life cycle energy consumption in the residential building, studied in the present work, varies between 5-10 % of its initial construction cost which corresponds at 59.84 €/m^2 to 123.84 €/m^2. The annual CO_2 emissions savings in the residential building have been estimated at 74 kgCO_2/m^2 to 134 kgCO_2/m^2. Therefore, creation of NZCEs buildings in Mediterranean region is technically and economically feasible using local benign green energy sources which are mature, reliable and cost-effective. Their construction requires development of appropriate policies and removal of various barriers which currently hinder their promotion. Further research should be oriented towards estimation of technical and economic feasibility of NZCEs buildings with nearly zero or zero net carbon emissions contributing in achieving carbon neutrality in Europe in the next decades.

References

[1] Suzuki, M., Oka, T. & Okada, K. (1995). "The Estimation of Energy Consumption and CO_2 Emission due to Housing Construction in Japan", *Energy and Buildings*, 22, pp. 165-169. https://doi.org/10.1016/0378-7788(95)00914-J

[2] Balaras, C.A., Gaglia, A.G., Georgopoulou, E., Mirasgedis, S., Sarafidis, Y. & Lalas, D.P. (2007). "European residential buildings and empirical assessment of the Hellenic building stock, energy consumption, emissions and potential energy savings", *Building and Environment*, 42, pp. 1298-1314.

[3] Poel, B., Cruchten, G.V. & Balaras, C.A. (2007). "Energy performance assessment of existing dwellings", *Energy and Buildings*, 39, pp. 393-403.

[4] Perez-Lombard, L., Ortiz, J. & Pout, C. (2008). "A review on buildings energy consumption information", *Energy and Buildings*, 40, pp. 394-398.

[5] Dulmage, S. & Mousa, M. (2015). "Embodied carbon white paper, Sustainable buildings Canada". Retrieved at 4/2/2020 from https://sbcanada.org/wp-content/uploads/2018/04/Embodied-Carbon-White-Paper.pdf

[6] Ramesh, T., Prakash, R. & Shukla, K.K. (2010). "Life cycle energy analysis of buildings: An overview", *Energy and Buildings*, 42, pp. 1592-1600. https://doi.org/10.1016/j.enbuild.2010.05.007

[7] Cellura, M., Guarino, F., Longo, S. & Mistretta, M. (2014). "Energy Life-Cycle Approach in Net Zero Energy Buildings Balance: Operation and Embodied Energy

of an Italian Case Study", *Energy and Buildings*, 72, pp. 371-381. https://doi.org/10.1016/j.enbuild.2013.12.046

[8] Berggren, B., Hall, M. & Wall, M. (2013). "LCE Analysis of Buildings—Taking the Step towards Net Zero Energy Buildings", *Energy and Buildings*, 62, pp. 381-391. https://doi.org/10.1016/j.enbuild.2013.02.063

[9] Vourdoubas, J. (2017). "Creation of zero CO_2 emissions residential buildings due to operational energy and embodied use on the island of Crete, Greece", *Open Journal of Energy Efficiency*, 6, pp. 141-154. DOI: 10.4236/ojee.2017.64011

[10] Ibn-Mohammed, T., Greenough, R., Taylor, S., Ozawa-Meida, L. & Acquaye, A. (2013). "Operational versus Embodied Emissions in Buildings—A Review of Current Trends", *Energy and Buildings*, 66, pp. 232-245. https://doi.org/10.1016/j.enbuild.2013.07.026

[11] Hernandez, P. & Kenny, P. (2008). "From net energy to zero energy buildings: Defining life cycle zero energy buildings", *Energy and Buildings*, 42, pp. 815-821. DOI: 10.1016/j.enbuild.2009.12.001

[12] Zhu, L., Hurt, R., Correa, D. & Boehm, R. (2009). "Comprehensive energy and economic analysis on a zero energy house versus a conventional house", *Energy*, 34, pp. 1043-1053.

[13] Hui, S. C. M. (2010). "Zero energy and zero carbon buildings: myths and facts", *In Proceedings of the International Conference on Intelligent Systems, Structures and Facilities (ISSF2010): Intelligent Infrastructure and Buildings*, 12 January 2010, Kowloon Shangri-la Hotel, Hong Kong, China, pp. 15-25.

[14] Cole, R.J. (2015). "Net-zero and net-positive design: a question of value", *Building Research & Information*, 43(1), pp. 1-6. doi: 10.1080/09613218.2015.961046

[15] Farhan, S.A., Shafiq, N., Mold Azizli, K.A., Jamaludin, N. & Safdar Gardezi, S.S. (2014). "Low carbon buildings: Renewable energy systems, materials and assessment methods", *Australian Journal of Basic and Applied Sciences*, 8(15), pp. 260-263.

[16] Iyer-Raniga, U. (2019). "Zero energy in the built environment: A holistic understanding", *Applied Science*, 9, 3375. doi:10.3390/app9163375

[17] Torcellini, P., Pless, S., Deru, M. & Crawley, D. (2006). "Zero energy buildings: A critical look at the definition", *Conference paper NREL /CP-550-39833, ACEEE summer study*, Pacific Grove, California, August 14-18, 2006.

[18] Marszal, A.J., Heiselberg, P., Bourrelle, J.S., Mussal, E., Voss, K., Sartori, I. et al. (2011). "Zero energy buildings—A review of definitions and calculation methodologies", *Energy and Buildings*, 43, pp. 971-979.

[19] Osmani, M. & O'Reilly, A. (2009). "Feasibility of zero carbon homes in England by 2016: a house builder's perspective", *Building and Environment*, 44 (9), pp. 1917-1924.

[20] Heffernan, E., Pan, W., Liang, X. & De Wilde, P. (2015). "Zero carbon homes: Perceptions from the EU construction industry", *Energy Policy*, 79, pp. 23-36. http://dx.doi.org/10.1016/j.enpol.2015.01.005

[21] Al-Temmamy, M.Z. & Abd-Rabo, L.M. (2019). "Sustainable design for zero carbon Architecture", *IOP conference series: Materials Science and Engineering*, 609. doi:10.1088/1757-899X/609/7/072001

[22] "Cost of carbon reduction in new buildings", Final report, Centre for sustainable energy, Currie and Brown, 2018. Retrieved at 4/2/2020 from https://www.bathnes.gov.uk/sites/default/files/sitedocuments/Planning-and-Building-Control/PlanningPolicy/LP20162036/cost_of_carbon_reduction_in_new_buildings_report_publication_version.pdf

[23] "Net zero carbon buildings: A framework definition", UK Green building council, 2019. Retrieved at 4/2/2020 from https://www.ukgbc.org/wp-content/uploads/2019/04/Net-Zero-Carbon-Buildings-A-framework-definition.pdf

[24] Vourdoubas, J. (2016). "Creation of zero CO_2 emissions residential building due to energy use: A case study in Crete, Greece", *Journal of Civil Engineering and Architecture Research*, 3(2), pp. 1251-1259.

[25] Xing, Y., Hewitt, N. & Griffiths, P. (2011). "Zero carbon buildings refurbishment – A hierarchical pathway", *Renewable and Sustainable Energy Reviews*, 15, pp. 3229-3236. doi:10.1016/j.rser.2011.04.020

[26] Luthander, R., Lingfors, D., Munkhammar, J. & Widen, J. (2015). "Self-consumption enhancement of residential photovoltaics with battery storage and electric vehicles in communities", *Proceedings of ECEEE summer study, Energy Use in Buildings, Projects, Technologies*, pp. 991-1002.

[27] Chandra Mouli, G.R., Bauer, P. & Zeman, M. (2016). "System design for a solar powered electric vehicle charging station for workplaces", *Applied Energy*, 168, pp. 434-443. http://dx.doi.org/10.1016/j.apenergy.2016.01.110

[28] "Smart EV charging systems for zero emission neighborhoods, A state-of-the-art study in Norway", (2018). Retrieved at 6/2/2020 from https://ntnuopen.ntnu.no/ntnu-xmlui/handle/11250/2504976

PART 2

[7] Creation of net zero CO$_2$ Emissions Hospitals Due to Energy Use. A Case Study in Crete, Greece

1. Introduction

Promotion of low energy and low CO$_2$ emissions buildings is at the core of current EU energy policy. Although hospital's energy systems are very complex using mainly conventional fuels the use of renewable energies is very important for them. Recent advances in various renewable energy technologies have increased their reliability and their cost effectiveness allowing their broad applications in new fields. Innovative financial tools like third party financing allow support of renewable energies investments by energy service companies (ESCOs). Recent changes in legislative framework regarding higher penetration of renewable energies in electric systems like attractive feed-in tariffs and net-metering regulations facilitate and promote green energy applications in new fields. Hospital buildings use large amounts of energy compared with other types of buildings and replacement of fossil fuels consumption with green energy sources is important for improvement of their environmental sustainability. Currently there are not many reports regarding the use of renewable energies in hospitals or studies concerning hospitals with low or zero CO$_2$ emissions due to energy use. Current work investigates the creation of hospitals with net zero carbon impacts due to energy use. Zero carbon emission hospitals contribute in climate change mitigation and in EU efforts to become a carbon neutral continent by 2050.

2. Literature survey

Use of geothermal energy in a hospital in New Zealand has been reported [Steins et al, 2012]. High enthalpy steam is used for covering the heating needs in the hospital while it could also be used for covering its future electricity requirements. The design of a solar absorption cooling system in a Greek hospital combined with a case study in Crete, Greece has been reported [Tsoutsos et al, 2010] while the cost effectiveness of this technology has been proved. An aquifer thermal storage system in a Belgian hospital has been reported [Vanhoudta et al,

2011]. The authors studied the storage system combined with a heat pump for heating and cooling in the hospital. They found that primary energy consumption of the heat pump was 71 % lower than a conventional system using gas-fired boilers and water cooling machines. The payback period of the investment without subsidies was estimated at 8.4 years. Energy consumption in Brazilian hospitals has been reported [Szklo et al, 2004]. For a medium size hospital the annual energy consumption varied between 230-460 KWh/m^2 depending on the comfort level while more electricity than heating fuel was used in it. Energy performance and energy conservation in health care buildings in Hellas has been reported [Santamouris et al, 1994]. The authors studied 30 health care buildings and they found that annual energy consumption varied from 407 KWh/m^2 in hospitals to 275 KWh/m^2 in clinics. Space heating required 73.4 % of the total energy consumption in hospitals and 65.3 % in clinics. They mentioned that proper energy saving measures could reduce their overall energy consumption by 10 %. Energy conservation techniques for hospital buildings have been reported [Kolokotsa et al, 2012]. The authors studied the state of the art technologies for energy efficiency in the hospital sector. They proposed that simple energy saving techniques can save up to 10 % of primary energy consumption. They also stated that a typical distribution of energy use in a good practice hospital with 500 beds is: fossil fuels for heating 34 %, fossil fuels for base load 14 %, lighting 14 % and in other electricity uses 38%. Use of renewable energies and energy conservation technologies in various buildings in Southern Europe has been reported [Santamouris et al, 1994]. They stated that the annual specific energy consumption per sector in a typical Greek hospital is: cooling 3 KWh/m^2, heating 299 KWh/m^2, lighting 52 KWh/m^2, electric equipment 53 KWh/m^2, total 407 KWh/m^2. Energy consumption in Greek hospitals has been reported [Sofronis et al, 2000]. The authors mentioned that the annual energy consumption varies according to the climate zone from 270 KWh/m^2 (south) to 438 KWh/m^2 (north). The average annual consumption was 370 KWh/m^2 shared at 290 KWh/m^2 for heating and at 80 KWh/m^2 for electricity. A report regarding saving energy in hospitals has been published [CADDET, 1997]. The report stated that annual energy consumption in hospitals varies between 250-1,000 KWh/m^2 while energy savings at 20-44 % can be achieved in them. It is also mentioned that annual thermal energy consumption in Greek hospitals is estimated at 300 KWh/m^2 while their annual electricity consumption at 110 KWh/m^2. An optimal operation of a complex power plant in a hospital generating energy for heating, power and cooling which can be profitable has been reported [Van Schijndel, 2002]. Heat consumption for preparing domestic hot water in hospitals has been studied [Bujak, 2010]. The author stated that in Poland during 2003-2008 the annual energy consumption in large hospitals with over 600 beds varied between 250-333 KWh/m^2. Use of new technologies for effective energy refurbishment in hospitals has been reported [Bizzari et al, 2006]. The authors studied the use of a phosphoric acid fuel cell, a solar thermal system and a solar-PV system in a hospital and concluded that their use in hospitals during refurbishment can be profitable. GHGs reduction and primary energy savings via the adoption of a fuel cell hybrid plant in a hospital has been reported [Bizzari et al, 2004]. The authors studied its use in a hospital in northern Italy and concluded that significant amounts of heat and electricity can be saved due to operation of the fuel cell. A study of different co-generation alternatives for a Spanish hospital has been published [Renedo et al, 2006]. The authors studied four co-generation systems providing heat, air conditioning and hot water in a large hospital in Spain and they concluded that all systems could be profitable while tri-generation in Southern Europe should be promoted. The economics of a tri-generation system in a hospital in Slovenia have been studied [Ziher et al, 2006]. The authors investigated the use of a tri-generation system with natural gas turbines and compression or absorption chillers for power, heat and cooling generation in a hospital and they concluded that the system can be profitable having low payback period and high net present value. The potential for incorporating low energy ventilation and cooling strategies into the design of new hospitals has been studied proving that they could decrease energy consumption and they could be profitable [Alan Short et al, 2009]. A study concerning energy use in Malaysian hospitals has been published [Saidur et al, 2010]. It was found that use of high efficiency electric motors can significantly decrease energy consumption achieving payback periods less than a year. Application of pinch technology in a hospital has been presented [Herrera et al, 2003] achieving power saving potential at 38 %. Energy consumption and indoor air quality in office and hospital buildings located in Athens, Greece has been studied [Argiriou et al, 1994]. The authors mentioned that the concentrations of NO_2 and SO_2 inside the hospital did not exceed the upper limits given by world health organization. A simple methodological tool for creation of buildings (hotels, dwellings) with net zero CO_2 emissions due to energy use has been presented, [Vourdoubas, 2015, Vourdoubas, 2015]. The author suggested the use of various renewable energies (solar thermal, solar-PV, solid biomass, geothermal energy with heat pumps) in buildings for covering all their energy needs in order to zero their net CO_2 emissions.

3. Energy consumption in hospitals

Energy is consumed in hospitals in various sectors including:
- Space heating,
- Space cooling,
- Hot water and steam production,
- Lighting, and
- Operation of various electric equipment and apparatus

Hospitals are among the highest energy consuming buildings and most of them have annual energy consumption at 200-400 KWh/m². Annual energy consumption in Greek hospitals according to various estimations (Santamouris et al, 1994, Caddett, 1997) varies between 407 KWh/m² and 410 KWh/m². Other estimations regarding energy consumption in hospitals vary according to the climate zone that the hospital is located. The main energy sources which are currently used in hospitals are heating oil and natural gas for heat production and electricity for lighting, air conditioning and operation of various electric equipment and devices. Distribution of energy consumption in a typical hospital per sector is presented in table 1

Table 1. Distribution of energy consumption in hospitals

Sector	Energy consumed (%) [1]	Energy consumed (%) [2]
Lighting	14	12.8
Other electricity uses	38	13.8
Heating	34	73.4
Other heat uses	14	
Total	100	100

[1]*Typical breakdown of energy use in a good practice hospital with 500 beds [Kolokotsa et.al., 2012],* [2] *Energy breakdown per sector in Greek hospitals [Santamouris et.al., 1994]*

Energy consumption depends on the quality of the building construction and its thermal insulation, on services offered as well as on local climate. Average low and high air temperatures in Chania, Crete, Greece vary between 7.9-14.1 °C in January and 21.2-30.6 °C in July. Solar irradiance in Chania at tilt 30 degrees varies between 83 KWh/m² in January to 208 KWh/m² in July with monthly average at 145 KWh/m². Fossil fuels and electricity used in hospitals can be easily replaced by renewable energies although there are not currently many hospitals using green energies in order to reduce their fossil fuels consumption. Application of various simple energy conservation techniques can save up to 10 % of primary energy consumption while the use of more sophisticated technologies can result in higher energy savings.

4. Use of renewable energies in hospitals

Although the energy system of a new hospital is extremely complex various renewable energies can be used in order to provide heat, cooling, electricity and vehicle's fuels in it [Renewable energy guide for European hospitals]. Among them are: a) various types of solid, liquid, or gaseous biomass, b) solar thermal energy, c) solar-PV energy, d) geothermal energy with heat pumps, e) wind turbines and hydropower systems if the hospital is located nearby a river or a waterfall and if the annual average wind velocity nearby the hospital is satisfactory. Wind energy and hydropower can be used according to their availability on-site or off-site. Biomass can be also used for heat or/and power generation. Additionally, bio-fuels can fuel hospital's vehicles. Biomass burning for heat generation is a mature, reliable and cost effective technology which is broadly used today. Solar thermal energy can be used also for heat (and cooling) generation. Currently it is broadly used for domestic hot water production in various buildings like residential buildings and hotels. Its effective use for space cooling has been proved only in large systems. Solar photovoltaic energy is also broadly used for power generation. The last few years the sharp decrease in solar cell prices has resulted in their increasing use for power generation mainly in grid connected systems. Ground source heat pumps are used broadly for heat and cooling generation in buildings. Heat pumps use electricity for their operation while they are energy efficient devices having C.O.P. in the range 3-4. The use of the abovementioned renewable energies in hospitals are influenced by various factors like: a) Their on-site or off-site availability, b) The existence of governmental supporting policies, favorable regulations and ambitious targets set, c) The existence of appropriate financial tools and mechanisms to support the necessary investments, d) The maturity, reliability and cost effectiveness of various renewable energy technologies which can be used in hospitals, and e) The possibility for grid connection of the electricity generation systems in the hospital. A guide towards zero carbon emissions hospitals using renewable energy technologies has been presented during the implementation of an IEE funded project "Res-hospitals" [http://www.res-hospitals.eu]. The case messages of this guide include the followings: Most European hospitals are embracing the need to reduce energy consumption but investments in renewable energy technologies are less common and very few of them are demonstrating financial innovation to support investments in renewable energies. Feasible options for on-site renewable energy production in European hospitals depend on various factors. However, the most popular choices for on-site renewable energy generation are biomass burning systems and photovoltaic panels but there is also an increasing number of ground source heat pumps used in them. Finally off-site energy generation systems including hydropower and wind farms offer the best opportunities in hospitals to exploit the benefits of renewable energies and to cope with the problems of rising energy costs.

5. Creation of hospitals with zero CO_2 emissions due to energy use

In order to zero net CO_2 emissions due to energy use in hospitals the following two criteria must be fulfilled.

A) Fossil fuels must not be used. They should be replaced by renewable energies,

B) Electricity used from the existing electric grid must be offset by green electricity generated by renewable energies like solar-PV electricity, and

C) Vehicles used in hospitals can be either electric vehicles with batteries or conventional vehicles with internal combustion engines. In the first case their electric batteries can be recharged with solar electricity while in the second they can use bio-fuels.

In this case net CO_2 emissions due to operating energy use in the hospital will be zero. Therefore, hospitals can produce heat using solar thermal energy and solid biomass, heat and cooling using ground source heat pumps while they can generate electricity with solar-PV panels. Combination of the abovementioned sustainable energy technologies can zero their net carbon emissions due to energy use achieving the target of zero net CO_2 emissions. Two cases are examined in the following sections for a hospital located in Chania, Crete, Greece.

5.1 Use of solar energy and solid biomass for covering all the energy needs in a hospital in Crete, Greece

In the first case the use of solar energy and solid biomass is examined. In this case the hospital will generate: a) domestic hot water mainly with solar thermal systems and additionally with solid biomass burning, b) space heating with solid biomass burning, and c) electricity with solar-PV panels. Electricity is used for lighting, space cooling and operation of various equipment and machineries (pumps, valves, lifts, aerators, refrigerators, surgical tools and equipment etc). The combined use of solar thermal energy, solar-PV energy and solid biomass in hospitals can zero their net CO_2 emissions due to energy use. In order to estimate the size of the necessary renewable energy systems the following assumptions are made:

a) The hospital has 300 beds and its covered area is 15,000 m² (50 m²/bed),
b) It will utilize bio-fuels instead of fossil fuels in its vehicles and renewable energies for heating, cooling and power generation,
c) A grid connected solar-PV system will generate annually the same amount of electricity that the hospital consumes from the grid (based on net-metering regulations),
d) The hospital will reduce by 10 % its initial annual energy consumption from 407 KWh/m² to 366 KWh/m² by implementing various energy saving measures,
e) The breakdown per sector of energy use in the hospital is 42% for electricity, 36% for space heating and other heat uses and 22% for hot water production,
f) The annual electricity generation from the solar-PV system in Crete is 1,500 KWh/KW$_p$,
g) The annual heat generation from the solar thermal system with flat plate collectors is 1,800 KWh/KW$_{th}$ (1 m² of flat plate solar collectors is equivalent at 0.7 KW$_{th}$),
h) The heat content of solid biomass is 4,000 kcal/kg while the efficiency of the biomass burning system is 0.75. Its net heating value is 3,000 kcal/kg (3.48 KWh/kg),
i) The flat plate collectors will produce 75% of the energy required for hot water production (906 MWh),
j) Solid biomass burning will produce all the energy used for space heating and other heat uses plus 25% of the required energy for hot water production (1,976 + 302 =2,278 MWh),
k) The power of the solid biomass burning system will be double than the average power estimated from annual needs (2,278,000 KWh)/(8,760 hours) =260X2=520KW) in order to cover the peak heating loads.

The energy consumption of the hospital is presented in table 2

Table 2. Energy consumption in the hospital

Number of beds	300
Covered area	15,000 m²
Specific annual energy consumption	366 KWh/m²
Total annual energy consumption	5,490 MWh
Electricity consumption	2,306 MWh
Energy used for hot water production	1,208 MWh
Energy used for space heating and other heat uses	1,976 MWh

The size of the solar thermal system, the solar-PV system and the solid biomass burning system is presented in table 3

Table 3. Size of various renewable energy systems covering all the annual energy needs in the hospital

Energy system	Size or Quantity consumed	Annual energy generation (MWh)
Nominal power of solar-PV system	1,537 KW$_{el}$	2,306
Area of flat plate collectors	503 m²	906
Power of flat plate collectors	352 KW$_{th}$	-
Power of solid biomass burning system	520 KW$_{th}$	2,278
Annual consumption of solid biomass	655 tons	-
Total		5,490

5.2 Use of solar energy and low enthalpy geothermal energy with heat pumps for covering all the energy needs in a hospital in Crete, Greece

In this case the hospital will generate:
a)   Hot water mainly with solar thermal systems and additionally with ground source heat pumps,
b)   Space heating and cooling with ground source heat pumps,
c)   Electricity with a solar-PV system which will be used for lighting and operation of various equipment and machineries (pumps, valves, lifts, aerators,

refrigerators, surgical tools and equipment etc). Additionally electricity will be also used for operation of the ground source heat pumps.

The solar thermal and the solar-PV system as well as the low enthalpy geothermal heat pump will utilize in-situ renewable energies. With reference to the hospital described previously the following assumptions are made in order to estimate the size of the necessary renewable energy systems.

a) The grid connected solar-PV will generate annually all the electricity required in the hospital,

b) The solar thermal system will produced 75% of hot water requirements (906 MWh),

c) The heat pump will cover all space heating requirements and other heating needs while it will produce 25% of the hot water needed (302 MWh). Additionally it will cover the cooling loads of the hospital while it is assumed that the necessary energy for that is equal at 15 % of the initial electricity consumption which corresponds at 346 MWh (1,976+302+346=2,624 MWh),

d) The C.O.P. of the heat pump is 3.5. Therefore, it will consume 750 MWh annually. It is assumed that it will operate 8,760 hours annually while its power will be double than the average estimated in order to cover peak loads. Therefore, the power of the heat pump is 750,000/(8, 760)=86X2= 172 KW,

e) Grid electricity consumption will be equal with the initial electricity consumption minus the electricity used for cooling which will be covered from the heat pump plus the electricity consumption of the heat pump (2,306-346+750 =2,710 MWh). The size of the renewable energy systems used in the hospital is presented in table 4.

Table 4. Size of various renewable energy systems which can cover all the energy needs in the hospital

Energy system	Size	Energy generation (MWh)
Area of flat plate collectors	503 m²	906
Power of flat plate collectors	352 KW$_{th}$	906
Power of heat pump	172 KW$_{el}$	2,624
Nominal power of solar-PV system	1,807 KW$_{el}$	2,710
Total		6,240

6. Economic Considerations

Use of the abovementioned renewable energy technologies in the hospital requires new investments. In order to estimate the necessary investment capital the following assumptions are made.

a) The cost of the solar thermal system is 250 €/m^2,
b) The cost of the solar-PV system is 1,200 €/KW$_p$,
c) The cost of the solid biomass burning system is 500 €/KW$_{th}$,
d) The cost of the ground source geothermal heat pump is 1,800 €/KW

The capital cost of the renewable energy systems which can generate all required energy in the abovementioned hospital in Crete, Greece is presented in table 5.

Table 5. Capital cost of various renewable energy systems which can generate all required energy in the hospital in Crete-Greece

Energy system	Capital cost of the solar energy and solid biomass systems (€)	Capital cost of the solar energy and geothermal heat pumps (€)
Solar thermal	125,750	125,750
Solar-PV	1,844,400	2,168,400
Solid biomass burning	260,000	-
Geothermal heat pump	-	309,600
Total	2,230,150	2,603,750
Total cost per bed	7,434	8,679

It is obvious that use of the abovementioned renewable energy technologies in the hospital reduces significantly the cost of conventional fuels which are necessary for covering all its energy needs. However, in order to assess the profitability of the proposed renewable energy systems their maintenance and depreciation cost must also be taken into account. The cost of solar and geothermal energy is zero and the only fuel cost is due to solid biomass use. The cost of conventional fuels as well as the cost of the renewable green fuels used is presented in table 6.

Table 6. Annual costs of fuels used for covering all the energy needs in a hospital in Crete-Greece [1]

Energy-fuel	Cost of conventional fuels (€/year)	Cost of solar energy and solid biomass (€/year)	Cost of solar energy and geothermal energy (€/year)
Grid electricity	576,500	-	-
Heating oil	304,689	-	-
Solid biomass	-	65,500	-
Total	881,189	65,500	0
Total per bed	2,937	217	0

[1] Cost of grid electricity, 0.25 €/KWh, Cost of heating oil, 1 €/kg, Cost of solid biomass (Olive kernel wood), 0.10 €/kg

7. Environmental benefits due to renewable energies use in a hospital

Use of various renewable energies in order to cover all energy requirements in a hospital in Crete, Greece results in environmental benefits due to savings in CO_2 emissions. In order to estimate CO_2 emission savings it will be assumed that the hospital uses conventional energy sources like electricity (generated from fossil fuels) and heating oil in order to cover all its heating needs. CO_2 emissions in the hospital resulted from the use of fossil fuels are presented in table 7.

Table 7. CO_2 emissions due to fossil fuels use in the hospital [1]

Fuel	Use	Annual consumption (MWh)	Annual CO_2 emissions (tons CO_2)
Electricity	Operation of various electric devices and machinery	2,306	2,281
Heating oil	Space heating, hot water production, other heat uses	3,184	975
Total		5,490	3,256
Total per bed		18.3	10.9

[1] Emissions coefficient for electricity, 0.989 kg CO_2/KWh, CO_2 emissions from heating oil, 3.2 kg CO_2/kg, Net heating value of heating oil, 9,000 kcal/kg (10.45 KWh/kg)

Therefore, the use of renewable energies covering all the energy needs of the abovementioned hospital located in Crete, Greece results in annual savings at 3,256 tonsCO_2.

8. Conclusions

Hospitals are high energy consuming buildings using mainly conventional energy sources like electricity, heating oil and natural gas. However, recent advances in renewable energy technologies allow their use in order to cover all the energy requirements in hospitals zeroing at the same time their net CO_2 emissions due to energy use. Although many hospitals are trying to use energy saving techniques and technologies in order to reduce their energy consumption there are not currently many applications of renewable energy technologies replacing fossil fuels use in them. Solar energy, solid biomass and low enthalpy geothermal energy are available in Mediterranean region and they can be used for power, heat and cooling generation in hospitals. A case study in a 300 beds hospital in Crete, Greece has shown that solar thermal energy, solar-PV energy and solid biomass can cover all its energy needs zeroing its net CO_2 emissions due to energy use. The necessary investments in the abovementioned renewable energy technologies have been estimated at 7,434 €/bed while the annual CO_2 emissions savings at 10.9 tonsCO_2/bed. An alternative case study in the same hospital has shown that combined use of solar thermal energy, solar-PV energy and low enthalpy geothermal energy with heat pumps can also cover all the energy requirements in the same hospital zeroing its net CO_2 emissions due to energy use. The necessary investments in that case are slightly higher than previously at 8,679 €/bed. Further work is needed in order to estimate the net present value of the abovementioned investments in green energies and assessing the profitability of using renewable energies instead of fossil fuels in hospitals.

References

[1] Alan Short, C. & Al-Maiyah, S. (2009). "Design strategy for low-energy ventilation and cooling of hospitals", *Building Research and Information*, 37, pp. 264–292.

[2] Argiriou, A., Asimakopoulos, D., Balaras, C., Daskalaki, E., Lagoudi, A., Loizidou, M., Santamouris, M. & Tselepidaki, I. (1994). "On the energy consumption and indoor air quality in office and hospital buildings in Athens Hellas", *Energy Conservation and Management*, 35, pp. 385-394.

[3] Bizzari, G. & Morini, G.L. (2006). "New technologies for an effective energy retrofit of hospitals", *Applied Thermal Engineering*, 26, pp. 161–169.

[4] Bizzari, G. & Morini, G.L. (2004). "Greenhouse gas reduction and primary energy savings via adoption of a fuel cell hybrid plant in a hospital", *Applied Thermal Engineering*, 24, pp. 383–400.

[5] Bujak, J. (2010). "Heat consumption for preparing domestic hot water in hospitals", *Energy and Buildings*, 42, pp. 1047–1055.

[6] Herrera, A., Islas, J. & Arriola, A. (2003). "Pinch technology application in a hospital", *Applied Thermal Engineering*, 23, pp. 127–139.

[7] Kolokotsa, D., Tsoutsos, Th. & Papantoniou, S. (2012). "Energy Conservation Techniques for Hospital Buildings", *Advances in Building Energy Research*, 6, pp. 159–172.

[8] Renedo, C.J., Ortiz, A., Manana, M., Silio, D. & Perez, S. (2006). "Study of different co-generation alternatives for a Spanish hospital center", *Energy and Buildings*, 38, pp. 484–490.

[9] Saidur, R. Hasanuzzaman, M., Yogeswaran, S., Mohammed, H.A. & Hossain, M.S. (2010). "An end-use energy analysis in a Malaysian public hospital", *Energy*, 35, pp. 4780-4785.

[10] Santamouris, M., Dascalaki, E., Balaras, C., Argiriou, A. & Gaglia, A. (1994). "Energy performance and energy conservation in health care buildings in Hellas", *Energy Conservation and Management,* 35, pp. 293-305.

[11] Santamouris, M. & Argiriou, A. (1994). "Renewable energies and energy conservation technologies for buildings in Southern Europe", *International Journal of Solar Energy*, 15, pp. 69-79.

[12] Sofronis, H. & Markogiannakis, G. (2000). "Energy consumption in public hospitals", *Bulletin of the Association of Mechanical Engineers*. Retrieved on 5/10/2015, from www.thelcon.gr/pdfs/publication%20hospitals.pdf. (in Greek).

[13] Steins, Ch. & Zarrouk, S.J. (2012). "Assessment of the geothermal space heating system at Rotorua Hospital, New Zealand", *Energy Conversion and Management*, 55, pp. 60–70.

[14] Szklo, A.S., Soares, J.B. & Tolmasquim, M.T. (2004). "Energy consumption indicators and CHP technical potential in the Brazilian hospital sector", *Energy Conversion and Management*, 45, pp. 2075–2091.

[15] Tsoutsos, Th., Aloumpi, E., Gkouskos, Z. & Karagiorgas, M. (2010). "Design of a solar absorption cooling system in a Greek hospital", *Energy and Buildings*, 42, pp. 265–272.
[16] Van Schijndel, A.W.M. (2002). "Optimal operation of a hospital power plant", *Energy and buildings*, 34, pp. 1055-1065.
[17] Vanhoudta, D., Desmedta, J., Van Baela, J., Robeynb, N. & Hoesb, H. (2011). "An aquifer thermal storage system in a Belgian hospital: Long-term experimental evaluation of energy and cost savings", *Energy and Buildings*, 43, pp. 3657–3665.
[18] Vourdoubas, J. (2015). "Creation of hotels with zero CO_2 emissions due to energy use: A case study in Crete-Greece", *Journal of Energy and Power Sources*, 2, pp. 301-307.
[19] Vourdoubas, J. (2016). "Creation of zero CO_2 emissions residential buildings due to energy use. A case study in Crete-Greece", *Journal of Civil Engineering and Architectural Research*, 3(2), pp. 1251-1259.
[20] Ziher, D. & Poredos, A. (2006). "Economics of a tri-generation system in a hospital", *Applied Thermal Engineering*, 26, pp. 680–687. Saving energy with Energy Efficiency in Hospitals, CADDET Energy Efficiency, CADDET, 1997.
[21] "Renewable Energy Guide for European Hospitals". Retrieved on 5/10/2015 from http://www.res-hospitals.eu
[22] "Towards Zero Carbon Hospitals with Renewable Energy Systems", IEE project report. Retrieved on 5/10/2015 from https://ec.europa.eu/energy/intelligent/projects/en/projects/res-hospitals

[8] Creation of zero CO_2 emissions school buildings due to energy use in Crete, Greece

1. Introduction

Improvement of energy performance in public and private buildings is obligatory in Europe as a necessary step for climate change mitigation and its transformation to carbon neutral continent by 2050. School buildings consume electricity for lighting and operation of various electric devices including space cooling and heat for space heating and hot water production. Their total energy consumption is low compared with other buildings like hospitals, residential buildings and offices. Current research investigates the possibility of zeroing carbon emissions in school buildings due to operating energy use focusing on the island of Crete, Greece. This can be achieved with the use of locally available renewable energy sources, like solar energy, solid biomass and low enthalpy environmental heat. Creation of net zero carbon emissions school buildings is important since it will indicate, as a good example, to students who are daily there the way that a building can minimize or zero its net carbon emissions due to energy use. Students can use the same methodology and green technologies in their residential buildings minimizing their carbon emissions. It is also important for achieving the national targets for GHG emissions reduction and climate change mitigation as well as the EU target of zeroing continent's carbon footprint by 2050.

2. Literature survey

Greek school buildings have total annual energy consumption at 68 KWh/m^2. Annual energy requirements for space heating have been estimated at 55 KWh/m^2 while for electricity at 13 KWh/m^2 [1]. Compared with other European school buildings they consume significantly less energy probably due to the mild climate in the country. The significant gap between design energy estimates and actual energy performance in school buildings has been investigated [2]. Results from a study in 15 school buildings across UK have shown that operational issues like occupants behavior have a major influence on their energy performance. Life cycle energy assessment in Australian secondary schools has been reported [3]. Embodied energy was estimated at 15.81-16.33 GJ/m^2 (4,395 KWh/m^2 - 4,540 KWh/m^2) while their operating energy at 21.53-40.52 GJ/m^2 (5,985 KWh/m^2 -

11,265 KWh/m^2) over a period of 60 years. Development of a methodology for energy performance benchmarking with reference to Irish primary schools has been presented [4]. Good practice guide for UK schools regarding annual heat consumption gives typical and best practice values for heating energy at 157 KWh/m^2 and at 110 KWh/m^2. Studies in 88 Irish schools estimated their annual median heating energy consumption at 96 KWh/m^2 with standard deviation at 50 KWh/m^2. An estimation model benchmarking the heating energy consumption in school buildings in Germany has been reported [5]. Annual mean heating energy consumption at 93 KWh/m^2 with standard deviation at 28 KWh/m^2 has been estimated while guidelines specify the average annual heating energy consumption for schools at 90 KWh/m^2. A benchmark on energy efficiency and GHG emissions in school buildings in central Argentina has been published [6]. Data analysis in 15 public school buildings has shown the average annual energy consumption at 122.7 KWh/m^2 with standard deviation at 41.1 KWh/m^2. Average annual CO_2 emissions in these educational buildings due to use of natural gas and electricity were estimated at 31.4 KgCO$_2$/m^2 with standard deviation at 10.5 KgCO$_2$/m^2. An analysis of energy consumption in high schools in a province in central Italy has been published [7]. The authors mentioned that thermal energy savings could reach at 38 % while electrical energy savings at 46 %. A review on buildings energy consumption has been presented [8]. The authors reported that in USA annual energy consumption in school buildings was at 262 KWh/m^2 which is substantially higher than in most E.U. countries. An assessment of heating energy consumption in 120 Italian school buildings in Torino area has been made and the average annual energy consumption in these buildings was estimated at 115 KWh/m^2 [9]. A report on energy performance in Hellenic schools has been presented [10]. Case studies in different schools which were implemented during 1970-2007 have shown that their annual heat energy consumption varied between 23.1-83.8 KWh/m^2 while their electric energy consumption between 7.7-28.55 KWh/m^2. A review of definitions and calculation methodologies of near zero energy buildings has been presented [11]. The authors stated that 60 % of the recorded indoor temperatures, one third of relative humidity and about 17-35 % of CO_2 concentration in the sample schools were inconsistent with indoor conditions prescribed by international standards. They have also mentioned that existing NZEB definitions and proposals for calculation methodologies indicated complexity of the concept and lack of common agreement in order to proceed in the deployment of NZEBs. However, NZEBs have the potential to significantly reduce energy use and to increase the overall share of renewable energies in the energy mix. Energy consumption and potential energy saving in 24 old school buildings in Slovenia has been studied [12]. The authors stated that their average annual energy consumption was at 192 KWh/m² the room air temperatures were too high while the indoor air quality was inadequate with CO_2 concentrations exceeding 4,000 p.p.m.. Use of display energy certificates to quantify school's energy consumption has been reported [13]. Taking into account a benchmark at 150 KWh/m² for annual thermal energy consumption and 40 KWh/m² for annual electrical energy consumption it was found that 45 % of primary schools were below the electrical energy benchmark and over 60 % of them below the thermal energy benchmark. Results from a big number of primary, secondary and academic schools indicated an average annual energy consumption of primary schools at 184 KWh/m² and annual CO_2 emissions at 52 $kgCO_2$/m². Average annual energy consumption and carbon emissions in secondary schools was at 188 KWh/m² and at 55 $kgCO_2$/m² while for academies it was at 205 KWh/m² and at 68 kg CO_2/m². The possibility of creating zero CO_2 emission buildings in Crete, Greece have been investigated [14], [15]. The author stated that renewable energies like solar thermal energy, solar-PV energy, solid biomass and low enthalpy geothermal energy combined with heat pumps can be used in order to cover all the energy requirements in various buildings replacing fossil fuels and zeroing their net CO_2 emissions due to energy use. The author stated that the abovementioned renewable energy technologies are reliable, mature and cost effective and their use can reduce substantially CO_2 emissions from buildings. Energy consumption in Greek schools is significantly lower than in similar schools in other countries according to existing literature. Apart from the mild climate in Greece, occupant's behavior, including proper controlling of indoor temperature influences their energy utilization resulting in lower energy consumption and additionally in lower CO_2 emissions. Investigation on benchmarking comparability regarding energy consumption in British schools has been presented [16]. Data analysis in 465 primary and secondary schools has shown that the average annual electricity consumption in secondary schools was at 51 KWh/m² slightly higher than in primary schools which annually consumed 44 KWh/m². Annual thermal energy consumption based on fossil fuels was almost the same at 122 KWh/m² and 121 KWh/m² respectively. A field study regarding energy consumption of school buildings in Luxemburg has been published [17]. Analyzing data from a large number of various schools the authors mentioned that the average annual thermal energy consumption was at 93 KWh/m² and the standard deviation was at 46 KWh/m². The fuels used in these buildings included gas, heating oil, district heat, wood pellets and co-generated heat with a gas boiler. They have also calculated the annual electricity consumption in 64 buildings separately where the mean value was estimated at 32 KWh/m^2 and the standard deviation at 15 KWh/m^2. A study regarding energy consumption in elementary schools in South Korea has been implemented [18]. The authors stated that the annual electricity consumption in 2010 was at 1,040 MJ/m^2 (289 KWh/m^2), oil consumption at 92 MJ/m^2 (25.6 KWh/m^2) and gas consumption at 325 MJ/m^2 (90 KWh/m^2). Results from various studies have shown that school buildings require more heat than electricity while their total annual energy consumption in Europe does not exceed 200 KWh/m^2 although it is influenced by many factors including the type of building construction, the local climate and occupant's behavior. However, in USA and in South Korea energy consumption in school buildings is higher than in Europe while two studies for Greek school buildings have indicated that their energy consumption is significantly lower than in other EU countries.

3. Energy consumption in school buildings in Greece

Annual energy consumption in Greek school buildings is low compared to school buildings in other countries. It is estimated at 55 KWh/m^2 for heating and at 13 KWh/m^2 for electricity, totally at 68 KWh/m^2. However, neither indoor temperatures nor the behavior of the occupants, students and teachers, were recorded in these estimations. The most common fuels used for heating were heating oil followed by gas. Solid biomass is not a popular fuel for heating school buildings in Greece. Electricity is used for lighting, operation of electric devices and for space cooling when needed. The energy consumption in a typical Greek school building as well as its CO_2 emissions due to energy use is presented in table 1.

Table 1: Energy consumption and CO_2 emissions in a typical Greek school building [1]

Energy	Annual energy consumption (KWh/m^2)	Annual CO_2 emissions (KgCO_2/m^2)
Heat	55	17.6
Electricity	13	10.4
Total	68	28.0

[1] *Electricity emission factor, 0.8 kgCO_2/KWh, Heating oil emission factor, 0.32 kgCO_2/KWh*

Total energy consumption in school buildings in Greece is low compared with consumption in other countries while total CO_2 emissions due to energy use are also lower. Annual carbon emissions in other countries have been estimated at 52-55 kgCO$_2$/m^2.

4. Use of renewable energies for energy generation in school buildings in Crete, Greece

Various renewable energy sources can be used in order to cover the energy requirements in schools. Among them are:

a) Solar energy including solar thermal and solar electricity
b) Solid biomass
c) Low enthalpy geothermal energy with ground source heat pumps

Their technologies are mature, reliable and cost effective and they can meet schools' requirements for electricity, heat and cooling. Although hot water requirements in schools are low, solar thermal systems can be used with simple thermosiphonic systems particularly in areas with high solar irradiance. The annual solar irradiance in Crete, Greece is at 1,700 KWh/m^2 to 1,880 KWh/m^2. Solar-PV energy can be used for electricity generation covering all the power needs in schools. When these systems are properly sized they can generate annually the same amount of electricity that school buildings consume. Taking into account net metering regulations in Greece solar-PV panels placed on the roof of grid connected school buildings can generate green electricity. Generated electricity is injected into the grid offsetting the grid electricity consumed. Current decrease in prices of solar-PV panels allows their increasing use in various buildings. Solid biomass is a cheap energy source compared to heating oil and gas in Greece while it is broadly used for space heating in various buildings. Many types of solid biomass like fire woods, wood pellets or wood briquettes can be used in Crete as heating fuel in schools. Biomass heating systems are operationally simple, cost effective and they can cover all heating requirements in school buildings. Low enthalpy geothermal energy with ground source heat pumps can be also used for heat and cooling generation in school buildings. These systems use electricity while they have a high coefficient of performance in the range 3-4 producing three to four times more heat and cooling compared with the electricity consumed. They have high installation cost but its operating cost is low resulting in their cost effectiveness over their life span. Geothermal heat pumps have currently increasing applications all over the world and they can be used for covering all heat and cooling requirements in school buildings in Crete.

5. Design of school buildings with zero CO_2 emissions due to energy use in Crete, Greece

According to EU directive 31/2010 on energy performance of buildings [19] all new public buildings after 31/12/2018 must be near zero energy buildings. Additionally, public buildings which undergo deep renovation must improve their energy performance. Therefore, use of fossil fuels in public buildings including school buildings must be decreased and the share of renewable energies in their energy mix must be increased. Buildings consume energy in their life cycle during construction, operation, refurbishment and demolition. Life cycle analysis in residential buildings [20] has shown that their embodied energy, which encompasses construction, refurbishment and demolition energy, corresponds at 10-20 % in total energy consumption while operating energy at 80-90 %. The approach followed in present work assumes that a public school building has decreased its energy consumption using various energy saving techniques and technologies while it uses only renewable energies for covering its energy requirements. The absence of fossil fuels consumption zeros its net CO_2 emissions due to energy use. Renewable energies used in the school building are located either on-site, like solar energy and geothermal energy, or off-site. In order to zero fossil fuels use and net CO_2 emissions in school buildings due to operating energy use the following two criteria must be fulfilled:

a)   *Fossil fuels must be replaced with renewable energies, and*

b)   *The grid electricity consumed in schools must be offset annually by on-site solar electricity generation and injected into the grid*

In the following analysis it has been assumed that all grid electricity is generated by fossil fuels. This is not true since part of it, which is currently estimated at 20 % in Crete-Greece, is generated by solar and wind energy. Therefore, the solar-PV system has been oversized. For heat production, instead of heating oil, either solid biomass or geothermal energy with ground source heat pumps can be used. For green electricity generation photovoltaic panels placed on the building's roof or terrace is the best option. Therefore, replacement of conventional fuels in school buildings can be achieved using the abovementioned sustainable energy technologies which can generate all required energy in them. The following two cases are examined.

a)   Solar photovoltaic use for electricity generation combined with solid biomass for heat production

b)   Solar photovoltaic use for electricity generation combined with geothermal heat pumps for heat and cooling production

5.1 Description of a school building which covers all its energy requirements with solar electricity and solid biomass in Crete, Greece

Assuming that specific energy consumption in the school building is the same as in table 1 and its covered surface is 1,000 m² the size of the solar-PV system and the biomass boiler are presented in table 2.

Table 2: Size and cost of renewable energy systems covering all the energy requirements in a school building located in Crete, Greece [1]

Heat required	55,000 KWh/year
Electricity required	13,000 KWh/year
Total energy required	68,000 KWh/year
Nominal power of the solar-PV system	8.7 KW$_p$
Cost of the solar-PV system	13,050 €
Power of solid biomass boiler	68.75 KW$_{th}$
Cost of solid biomass boiler	34,375 €
Quantity of solid biomass required	15.8 tons/year
Cost of biomass use during operation	1,580 €/year
Total capital cost of both renewable energy systems	47,425 €

[1] *Operation of the heating system, 800 h/year, Net calorific value of solid biomass, 3,000 kcal/kg, Cost of solid biomass in Crete, 0.1 €/kg, Unit costs: Biomass boiler, 500 €/KWth, Solar-PV system: 1,500 €/KWp*

5.2 Description of a school building which covers all its energy requirements with solar electricity and a geothermal heat pump in Crete, Greece

Alternatively, energy requirements in the school building can be covered with a solar-PV system providing the electricity needed and a geothermal heat pump covering its heating and cooling needs. The heat pump is sized to cover the peak heating load which is assumed to be 50% higher than the average load. Since the heat pump uses electricity the size of the solar-PV system will be higher compared with the previous case since it will generate electricity for lighting, operation of various electric devices and additionally the electricity needed by the heat pump. Assuming that the specific energy consumption in the school building will be the same as in table 1 and its covered surface is 1,000 m² the size and the cost of the solar-PV system and the geothermal heat pump are presented in table 3.

Table 3. Size and cost of renewable energy systems covering all energy requirements in a school building in Crete, Greece [1]

Heat required	55,000 KWh/year
Electricity required	13,000 KWh/year
Total energy required	68,000 KWh/year
Additional electricity required for the operation of the heat pump	15,714 KWh
Total electricity required including the consumption of the heat pump	28,714 KWh
Nominal power of the solar-PV system	19.14 KW$_p$
Cost of the solar-PV system	28,710 €
Operation of the heat pump	800 h/year
Coefficient of performance of the heat pump	3.5
Power of the heat pump	29.5 KW
Cost of the heat pump	59,000 €
Total capital cost of renewable energy systems	87,710 €
Cost of energy/fuels used during operation	0 €/year

[1] *Heat pump is sized at 150 % of the base load, Unit costs: Solar-PV system, 1,500 €/KW$_p$, Geothermal heat pump, 2,000 €/KW*

Comparing tables 3 and 4 it is concluded that the size of the solar-PV system is higher in the second case while the capital cost of both required green energy systems is also higher than in the first case. However, the annual operating cost of fuels used in the second case is zero compared with 1,580 €/year in the first case although these values will be different if maintenance and depreciation costs are taken into account.

6. Economic considerations

Use of various renewable technologies in school buildings is currently cost effective and installation of renewable energy systems can be easily justified. Capital cost estimations for the abovementioned renewable energy systems installed in a school building with covered area at 1,000 m² are presented in table 4.

Table 4. Capital cost estimation of renewable energy systems covering all the annual energy needs in a school building located in Crete, Greece

Cost	Use of a solid biomass heating system and solar-PV panels	Use of a geothermal heat pump and solar-PV panels
Cost of solar-PV (€)	13,050	28,710
Cost of solid biomass burning system (€)	34,375	-
Cost of geothermal heat pump (€)	-	59,000
Total capital cost (€)	47,425	87,710
Total capital cost per m^2 of covered area in the school building (€/m^2)	47.42	87.71

It is concluded that in the second case when the school building is heated and cooled with the geothermal heat pump the total capital cost of renewable energy systems is significantly higher than in the first case of heating it with solid biomass.

7. Environmental and social considerations

Estimation of CO_2 emissions savings due to renewable energy use in school buildings with a covered area of 1,000 m^2 in Crete, Greece are presented in table 5.

Table 5: CO_2 savings in school buildings which use renewable energies to cover all their energy requirements in Crete, Greece

Energy systems	Capital cost of renewable energy systems (€)	Annual CO_2 savings (KgCO$_2$)	Capital cost of renewable energy systems per annual CO_2 savings (€/kgCO$_2$)
Use of solar-PV and solid biomass	47,425	28,000	1.69
Use of solar-PV and geothermal heat pumps	87,710	28,000	3.13

Use of renewable energies instead of fossil fuels in school buildings is important in Greece which is highly depended on imported fossil fuels. Additionally, it results in benefits to local economy due to use of local benign energy resources and renewable energy systems which can be constructed and maintained locally. Students and teachers will learn about sustainable energy technologies and their environmental benefits since the renewable energy systems installed in school buildings will act as a living laboratory where they spend many hours daily. Probably many of them will try to install these green energy technologies in their homes.

8. Conclusions

School buildings consume less energy than other types of buildings while Greek schools consume significantly less heat and electricity than the same buildings in other countries. Annual heat requirements in school buildings in Greece are estimated at 55 KWh/m^2 while electricity requirements at 13 KWh/m^2. Total annual CO_2 emissions are estimated at 28 kgCO_2/m^2. Changes in EU and national legislation require the decrease of energy consumption and the increase of the share of renewable energies in public buildings, including school buildings, in the near future. School buildings consume energy mainly for heating, cooling, lighting and operation of various electric devices. Current advances in various renewable energy technologies allow their use in a cost effective way in school buildings replacing the use of fossil fuels in them. Solar energy, solid biomass and ground source heat pumps can be used in school buildings in Crete, Greece for covering all their annual energy requirements zeroing net CO_2 emissions in them. Use of the abovementioned renewable technologies is cost effective and apart from economic benefits their use results in environmental and social benefits as well. The total capital cost of the required benign energy systems, in order to zero net CO_2 emissions in Greek school buildings, has been estimated at 47.42 €/m^2 to 87.71 €/m^2 while the annual carbon savings at 1.69 €/kgCO_2 to 3.13 €/kgCO_2. In order to verify these cost estimations and to prove the technical, economic and operational feasibility of the proposed sustainable energy systems their installation in few pilot school buildings in Crete, Greece is suggested in the future.

References

[1] "Report on Energy saving in Greek buildings" (2008). University of Athens, (In Greek). Retrieved on 25/1/2016 from http://www.sate.gr/nea/energy.pdf

[2] Demanuele, C., Tweddell, T. & Davies, M. (2010). "Bridging the gap between predicted and actual energy performance in schools", In *World renewable energy congress: World renewable energy congress and exhibition; WREC XI*: Abu Dhabi, United Arab Emirates, 2010. Brighton: WREC

[3] Ding, G. K. C. (2007). "Life cycle energy assessment of Australian secondary schools", *Building Research & Information*, 35(5), pp. 487–500. doi:10.1080/09613210601116408

[4] Hernandez, P., Burke, K. & Lewis, J. O. (2008). "Development of energy performance benchmarks and building energy ratings for non-domestic buildings: An example for Irish primary schools", *Energy and Buildings*, 40(3), pp. 249–254. doi:10.1016/j.enbuild.2007.02.020

[5] Beusker, E., Stoy, C. & Pollalis, S. N. (2012). "Estimation model and benchmarks for heating energy consumption of schools and sport facilities in Germany", *Building and Environment*, 49, pp. 324–335. doi:10.1016/j.buildenv.2011.08.006

[6] Filippín, C. (2000). "Benchmarking the energy efficiency and greenhouse gases emissions of school buildings in central Argentina", *Building and Environment*, 35(5), pp. 407–414. doi:10.1016/s0360-1323(99)00035-9

[7] Desideri, U. & Proietti, S. (2002). "Analysis of energy consumption in the high schools of a province in central Italy", *Energy and Buildings*, 34(10), pp. 1003–1016. doi:10.1016/s0378-7788(02)00025-7

[8] Pérez-Lombard, L., Ortiz, J. & Pout, C. (2008). "A review on building's energy consumption information", *Energy and Buildings*, 40(3), pp. 394–398. doi:10.1016/j.enbuild.2007.03.007

[9] Corgnati, S. P., Corrado, V. & Filippi, M. (2008). "A method for heating consumption assessment in existing buildings: A field survey concerning 120 Italian schools", *Energy and Buildings*, 40(5), pp. 801–809. doi:10.1016/j.enbuild.2007.05.011

[10] Dascalaki, E. G. & Sermpetzoglou, V. G. (2011). "Energy performance and indoor environmental quality in Hellenic schools", *Energy and Buildings*, 43(2-3), pp. 718–727. doi:10.1016/j.enbuild.2010.11.017

[11] Marszal, A. J., Heiselberg, P., Bourrelle, J. S., Musall, E., Voss, K., Sartori, I. & Napolitano, A. (2011). "Zero Energy Building – A review of definitions and

calculation methodologies", *Energy and Buildings*, 43(4), pp. 971–979. doi:10.1016/j.enbuild.2010.12.022

[12] Butala, V. & Novak, P. (1999). "Energy consumption and potential energy savings in old school buildings", *Energy and Buildings*, 29(3), pp. 241–246. doi:10.1016/s0378-7788(98)00062-0

[13] Godoy-Shimizu, D., Armitage, P., Steemers, K. & Chenvidyakarn, T. (2011). "Using Display Energy Certificates to quantify schools' energy consumption", *Building Research and Information*, 39(6), pp. 535-552. http://dx.doi.org/10.1080/09613218.2011.628457

[14] Vourdoubas, J. (2015). "Creation of zero CO_2 Emissions Hospitals Due to Energy Use. A Case Study in Crete-Greece", *Journal of Engineering and Architecture*, 3(2), pp. 1-9.

[15] Vourdoubas, J. (2016). "Creation of Zero CO_2 Emissions residential Buildings due to Energy Use: A Case Study in Crete-Greece", *Journal Civil Engineering Architecture Research*, 3(2), pp. 1251-1259.

[16] Hong, S.-M., Paterson, G., Mumovic, D. & Steadman, P. (2013). "Improved benchmarking comparability for energy consumption in schools", *Building Research & Information*, 42(1), pp. 47–61. doi:10.1080/09613218.2013.814746

[17] Thewes, A., Maas, S., Scholzen, F., Waldmann, D. & Zürbes, A. (2014). "Field study on the energy consumption of school buildings in Luxembourg", *Energy and Buildings*, 68, pp. 460–470. doi:10.1016/j.enbuild.2013.10.002

[18] Tae-Woo, K., Kang-Guk, L. & Won-Hwa, H. (2012). "Energy consumption characteristics of the elementary schools in South Korea", *Energy and Buildings*, 54, pp. 480–489. doi:10.1016/j.enbuild.2012.07.015

[19] Directive 2010/31/EU of the European Parliament and of the Council of 19 May 2010 on the energy performance of buildings (recast) [2010] OJ L153/13

[20] Ramesh, T., Prakash, R. & Shukla, K. K. (2010). "Life cycle energy analysis of buildings: An overview", *Energy and Buildings*, 42(10), pp. 1592–1600. doi:10.1016/j.enbuild.2010.05.007

[9] Possibilities of creating swimming pools with zero CO_2 emissions due to energy use. A case study in Crete, Greece

1. Introduction

Mitigation of climate change requires the sharp decrease in fossil fuels use replacing them with various renewable energies. Europe has declared its decision to zero its carbon emissions by 2050 eliminating its carbon footprint. Swimming pools utilize energy for covering their needs in water heating, hot water use in bathrooms as well as in electricity. The purpose of the current study is to examine the possibility of using a known methodology for zeroing net CO_2 emissions in swimming pools due to operating energy use. This can be achieved with the use of reliable, mature and cost-effective renewable energy technologies instead of conventional fossil fuels which are currently used. A case study analysis in a swimming pool located in Crete, Greece demonstrates the proposed methodology and the possibility of using various locally available renewable energy resources, with appropriate technologies, providing heat and electricity in it. Creation of private and public carbon neutral swimming pools due to energy use is important for reducing GHG emissions, mitigating climate change and creating a low carbon economy in Europe and worldwide.

2. Literature review

Creation of a low carbon economy consists of one of the main targets of European Union. According to EU directive 2009/28/EC by 2020 GHG emissions must be reduced by 20 %, 20 % of final energy consumption must be covered by renewable energy resources and energy efficiency must be increased by 20 %. It is expected that transition to a low carbon economy worldwide will result in anthropogenic temperature rise less than 2°C (Carvalho et al, 2011). Buildings use up to 40 % of total energy consumption in E.U. while the use of renewable energies in buildings, replacing fossil fuels, is easier than in other sectors of economy like industry, agriculture and transport (Urge-Vorsatz et al, 2007). GHG emissions in buildings can be reduced with various ways including:
a) Reducing energy consumption in them,
b) Replacing fossil fuels with renewable fuels, and
c) Using energy efficient technologies having low carbon emissions

Swimming pools utilize energy for space heating and ventilation, for water heating, for lighting and for the operation of various electric devices (Energy efficiency in swimming pools). Energy cost consists at 25-30 % of their total operating cost. Average pool water temperatures in indoor swimming pools vary between 26-30 °C. Measures improving energy efficiency in swimming pools include:

a)   Use of co-generation of heat and power systems,
b)   Use of condensing boilers,
c)   Heat recovery from the ventilation system, and
d)   Heat recovery from the pool water

According to the same study the distribution of energy use in indoor swimming pools is 53 % for space heating, 25 % for water heating, 6.5 % for lighting and 15.5 % for the operation of electric equipments. Therefore, indoor swimming pools use more energy in heating than in operation of various electric apparatus and lighting. Indoor swimming pools utilize significantly more energy per m^2 than various other buildings like residential buildings, offices, hospitals, schools, etc. Trianti-Stourne et al, 1998 have studied energy conservation strategies in swimming pools. The authors stated that for Mediterranean type climate the average annual total energy consumption per water pool surface area is at 4,300 KWh/m^2. For the continental European climate the corresponding energy consumption is higher at 5,200 KWh/m^2. Space and water heating as well as ventilation are important energy consumption sectors. According to the authors typical energy consumption in swimming pool facilities is made up of 45 % for ventilation of the pool hall, 33 % for pool water heating, 10 % for heat and ventilation for the remainder of the building, 9 % for electricity used in electric equipment and lighting and 3 % for hot water production in bathrooms. A report on energy efficiency in swimming pools has been presented by Kampel, 2015. The author has collected data from 43 Norwegian swimming facilities and has analyzed them. He found a significant variation in final annual energy consumption and he estimated that the potential reduction of their energy use could be approximately at 28 %. Analyzing a large number of data concerning annual energy consumption in swimming pools he estimated the average value at 4,004 KWh/m^2 with a standard deviation at 1,821 KWh/m^2. He also mentioned that various studies in Scandinavian countries have estimated the annual energy consumption in swimming pools. In Sweden it varies between 1,500-8,400 KWh/m^2 with an average value at 4,481 KWh/m^2 and in Denmark between 2,291-2,608 KWh/m^2. Reports from Finland estimated the annual energy consumption

at 4,475 KWh/m^2. Data on the distribution of energy use in swimming pools vary considerably. However, the highest amount of energy is consumed for space and water heating. Kampel, 2015 has also reported that according to British swimming pools association 52 % of energy is used in ventilation, 26 % for water heating, 5 % for lighting and the rest 17 % for operation of various electric devices. Vourdoubas, 2015 and Vourdoubas, 2016 has presented a methodology for zeroing net CO_2 emissions in various buildings due to operating energy consumption using various renewable energy sources and technologies. He stated that the current legal framework including net-metering regulations, their technological improvements and the cost decrease in various renewable energy technologies allow their broad application in buildings for heat and power generation. According to him in order to zero the net CO_2 emissions due to energy use in buildings two criteria must be fulfilled. The first is to replace all fossil fuels used with renewable energies and the second is to offset annually the grid electricity use with solar-PV electricity. Various renewable energy technologies could be used in a building in order to zero its net CO_2 emissions. The author presented various case studies for buildings located in Crete, Greece. The buildings
used solar thermal energy, solid biomass and low enthalpy geothermal energy with heat pumps for providing heat, cooling and green electricity. Current technological advances in the abovementioned renewable energy technologies have decreased their cost and have increased their reliability allowing their use in buildings for covering all their energy needs.

3. Use of renewable energies in swimming pools

Solar thermal energy is the most commonly used green energy in swimming pools. A report on solar energy use in outdoor swimming pools (SOLPOOL report, 2008) stated that use of solar heating systems in outdoor swimming pools varies significantly among EU countries. According to this report there are approximately 6,700 public swimming pools in Germany while half of them are outdoors pools. In the end of 2007 there were 799 public pools in Germany equipped with solar thermal systems. The absorber surface area of the installed solar thermal systems is greater than 100 m^2. In other EU countries the number of installed solar thermal systems in swimming pools is smaller. In Greece the solar irradiance is high and the number of swimming pools is also high. However, only a small percentage of them is equipped with water heating systems. During July and August when local temperatures are high outdoors pools do not require heating. A method for solar heating in outdoor swimming pools by placing a plastic cover on the water surface in Australia has been described by Czarnecki, 1963. Mathematical modeling and simulation of thermal performance of a solar heated indoor swimming pool has been reported (Mancic et al, 2014). The authors stated that up to 87 % of water heating demand can be met by solar heating systems. Use of geothermal energy in swimming pools in Island and Ecuador has been reported by Haraldsson et al, 2014. The authors stated various applications of direct use of geothermal fluids with temperatures between 45-75°C for heating swimming pools in Ecuador. Ground source heat pumps were also used for space heating in touristic complexes. In Iceland, they reported, there are many indoor and outdoor swimming pools utilizing hot geothermal fluids for water heating. The feasibility of heating residential swimming pools with geothermal heat pumps in USA has been reported by Chiasson. The author stated that in northern regions of USA it is not economically feasible to heat pools with ground source heat pumps (GSHP). On the contrary in southern areas it is economically justified to heat pools with GSHP. An analysis of heat pump's applications in large public buildings in China has been presented by Liu et al, 2015. The authors mentioned the advantages and drawbacks of heat pumps used in this country. An analysis of ground source heat pumps used in Cyprus has been presented by Michopoulos et al, 2015. The authors stated that significant economic and environmental benefits can be achieved with the substitution of conventional heating systems with ground source heat pumps in the country. Biomass can be used for space and water heating in swimming pools. In general biomass use is important when 100 % renewable energy systems are going to be used for energy supply. Studies for 100 % renewable systems in countries like Denmark (Lund et al, 2009), Ireland (Connolly et al, 2011) and Macedonia (Cosic et al, 2012) have been reported. Various combinations of renewable energy systems use were proposed in order to ensure that countries are going to be independent from fossil fuels. In all cases, among renewable energies, biomass is an important energy source for heat generation, electricity generation and production of transportation fuels. The limitations regarding biomass use for heating in 100 % renewable energy systems has been presented by Mathiesen et al, 2012. Since biomass is an important energy source for heating in these systems the authors presented alternative renewable energy sources which can cover the heating loads allowing use of biomass in other sectors like power generation. The possibility of using biomass for energy generation replacing fossil fuels has been presented by Balat et al, 2006. The authors stated that biomass can be considered as carbon neutral fuel assuming that biomass stock is not gradually diminished. Wood biomass as sustainable energy source for greenhouse heating in Italy has been reported by Bibbiani et al, 2016. The authors stated that solid biomass is highly recommended regarding its economic feasibility. They also reported that it can be considered as carbon neutral green fuel excluding GHG emissions generated during its harvesting, processing and transportation to the consumption site. The possibility of using solar-PVs in zero energy buildings has been reported by Scognamiglio et al, 2012. The authors stated that after 2020 solar-PVs will play an important role in the implementation of near zero energy buildings in Europe. Investigation of using various solar energy technologies in order to achieve a near zero energy building has been reported by Good et al, 2015. The authors studied various solar energy systems including solar thermal, solar-PV and combined solar thermal/PV stating that the best performance was achieved with the solar-PV system. The improvement of the renewable energy mix in order to achieve zero energy status in a building has been presented by Visa et al, 2014. The authors mentioned a case study of a solar house which used a combination of geothermal energy, solar thermal and solar-PV energy for achieving its zero energy status. The abovementioned study indicated that various renewable energy sources can be used for covering all the energy needs in swimming pools zeroing their net CO_2 emissions due to energy use. 100 % renewable swimming pools are technologically and economically feasible since the required green energy technologies are mature, reliable and cost-effective. Solar thermal energy and solid biomass could be used for heat generation. Ground source heat pumps can be also used for heat and cooling production. Space cooling is needed in pool's hall. Finally, solar-PV energy can be used for electricity generation. The necessary solar-PV system could be sized in order to generate the same amount of electricity that the pool consumes all over the year according to net metering regulations. The possibility of using these green technologies in swimming pools is presented in table 1.

Table 1. Use of renewable energy technologies in swimming pools.

Energy source	Energy generation	Possibility of covering all the needs in heat or/and electricity
Solar thermal	Heat	No
Solid biomass	Heat	Yes
Low enthalpy geothermal energy with heat pumps	Heat and cooling	Yes

| Solar-PV | Electricity | Yes |

Source: Own estimations

4. Creation of zero CO_2 emission swimming pools due to energy use

In order to zero net CO_2 emissions during operation in swimming pools the following two criteria must be fulfilled:

a) The swimming pool must use only renewable energy sources for covering all its heating needs replacing fossil fuels, and

b) It must generate all the grid electricity consumed annually with solar-PV electricity according to net-metering regulations. In this case the pool offsets grid electricity use based on fossil fuels with green solar electricity.

It has been assumed that various renewable energies used, including solid biomass, do not produce GHGs while grid electricity is generated from fossil fuels. When the previous two criteria are fulfilled it is concluded that the carbon footprint due to energy use in the swimming pool is zero. Among renewable energy sources which could be used for space and water heating in swimming pools, solid biomass, geothermal energy with heat pumps and solar thermal energy have been already used in many applications in buildings and in other sectors. Their technologies are reliable, mature and cost effective. In the case of solid biomass it is assumed that its net CO_2 emissions are zero neglecting any emissions during its harvesting, processing and transportation.

5. Creation of a zero CO_2 emissions swimming pool. A case study in Crete, Greece

In order to size and assess the energy systems used in a zero CO_2 emissions swimming pool located in Crete, Greece the following assumptions are made.

a) The pool's water surface area is 750 m^2,

b) The annual energy consumption in the pool is 4,000 KWh/m^2,

c) The annual heat consumption in the pool is 3,400 KWh/m^2,

d) The annual electricity consumption in the pool is 600 KWh/m^2, and

e) The annual electricity generation from a solar-PV system in Crete, Greece is 1,500 KWh/KW_p

Two different cases are examined as follows:

In the first case solar thermal energy and solid biomass will be used equally for covering all its heating needs while in the second solar thermal energy and ground source heat pumps are also equally used for that. In both cases solar-PV panels will generate all the required electricity.

5.1 Use of solar thermal energy and solid biomass for heat generation

Flat plate solar collectors will be used for heat generation as well as solid biomass. The annual heat consumption in the pool is 2,550 MWh$_{th}$ equally provided from solar energy and biomass. It is assumed that the annual heat generation of solar collectors in Crete, Greece is 1.2 MWh/m^2 (Tsoutsos et al, 2009), the heat content of solid biomass is 4,200 Kcal/kg and the efficiency of the biomass burning system is 75 %. The price of solid biomass is 150 €/ton. The biomass boiler will operate 2,550 hours annually and its power will be 500 MWh$_{th}$. The annual electricity consumption in the swimming pool is 450 MWh$_{el}$. CO_2 emission coefficients for heating oil is 0.32 kgCO$_2$/KWh and for grid electricity 0.75 kgCO$_2$/KWh. The sizing of the necessary renewable energy systems in order to generate all the energy required in the swimming pool is presented in table 2. Their capital cost as well as their operating cost is also presented in the same table.

Table 2. Size and cost of the green energy systems

Annual heat consumption in the pool	2,550 MWh$_{th}$
Heat provided from solar collectors	1,275 MWh$_{th}$
Size of flat plate solar collectors	1,062.5 m^2
Cost of flat plate collectors (300 €/m^2)	318,750 €
Heat generated from solid biomass	1,275 MWh$_{th}$
Power of the biomass burning system	500 MW$_{th}$
Cost of the biomass burning system (250 €/KW$_{th}$)	125,000 €
Annually required biomass	348.5 tons
Cost of annually required biomass during its operation	52,275 €
Annual electricity consumption in the pool	450 MWh$_{el}$
Nominal power of the solar-PV system	300 KW$_p$
Cost of the solar-PV system (1,200 €/KWp)	360,000 €
Annual cost of renewable fuels during its operation	52,275 €
Total capital cost of the required renewable energy systems	803,750 €

In order to estimate the annual CO_2 emission savings due to use of renewable energies it is assumed that in the case that the swimming pool was using heating oil for heat generation and grid electricity for lighting and powering various electric apparatus its CO_2 emissions are:

a) Due to heating oil use, 2,550 MWh X 320 kgCO$_2$/MWh = 816,000 kgCO$_2$
b) Due to grid electricity use, 450 MWh X 750 kgCO$_2$/MWh = 337,500 kgCO$_2$

Therefore, its total carbon emissions are: 1,153,500 kgCO$_2$/year

In order to estimate the quantities and costs of the conventional fuels used, including heating oil and grid electricity, for covering all the energy needs in the swimming pool it is assumed that the net heating value of heating oil is 11 KWh/kg, its cost is 1 €/kg and the cost of grid electricity is 0.25 €/KWh. The required annual quantity of heating oil is 232 tons, its cost is 232,000 € while the cost of electricity is 112,500 €. Therefore, the overall annual cost of conventional fuels used for providing all the required energy in the pool is 344,500 €.

5.2 Use of solar thermal energy and a ground source heat pump for heat generation

Flat plate solar collectors will be used for heat generation together with a ground source heat pump. The annual heat consumption in the pool is 2,550 MWh$_{th}$ equally provided from solar and geothermal energy. The GSHP will operate 2,550 hours per year and its C.O.P. is 4.5. Its power will be at 111 KW$_{el}$. The sizing of the necessary renewable energy systems including the solar thermal system, the solar-PV and the GSHP in order to generate all the energy required in the swimming pool is presented in table 3. The capital cost of these systems has been also estimated.

Table 3. Size of renewable energy systems covering all energy needs in a swimming pool located in Crete, Greece

Annual heat consumption in the pool	2,550 MWh$_{th}$
Heat provided from solar collectors	1,275 MWh$_{th}$
Size of flat plate solar collectors	1,062.5 m^2
Cost of flat plate collectors (300 €/m^2)	318,750 €
Heat generated from the GSHP	1,275 MWh$_{th}$
C.O.P. of the GSHP	4.5
Power of the GSHP	111 KW$_{el}$, 500 KW$_{th}$
Cost of the GSHP (1,800 €/KW$_{el}$)	200,000 €
Electricity consumption of the GSHP	283.05 MWh$_{el}$
Annual electricity consumption in the pool excluding the consumption of the GSHP	450 MWh$_{el}$
Total electricity consumption in the pool	733.05 MWh$_{el}$

Nominal power of the solar-PV system	488.7 KW$_p$
Cost of the solar-PV system (1,200 €/KW$_p$)	586,440 €
Annual cost of renewable fuels providing all the required energy	0 €
Total capital cost of the required renewable energy systems	1,105,190 €

5.3 Comparison of two different combinations of renewable energy systems providing all the required energy in the pool

The results of analysis in the abovementioned different combinations of renewable energy systems generating all the required energy in the pool are presented in table 4. Ignoring the depreciation and maintenance cost of the renewable energy systems the payback period of the green energy investments has been also estimated assuming that these energy systems will eliminate all the conventional fuels used in it. The payback period of the investments has been estimated as the ratio of capital investments to annual cost savings due to lower fuel costs.

Table 4. Comparison of two different combinations of renewable energy systems which could provide all the required energy in a swimming pool in Crete, Greece

Total capital cost (solar energy and biomass)	803,750 €
Total capital cost per m² of the pool water surface	1,072 €
Total capital cost (solar energy and GSHP)	1,105,190 €
Total capital cost per m² of the pool water surface	1,474 €
Operating cost (solar energy and biomass)	52,275 €
Decrease of the operation cost in the pool due to installation of renewable energy systems (solar energy and biomass)	292,225 €
Operating cost (solar energy and GSHP)	0 €
Decrease of the operation cost in the pool due to the installation of renewable energy systems (solar energy and GSHP)	344,500 €
Annual CO$_2$ savings due to the use of renewable energy systems	1,153,500 kg CO$_2$
Annual CO$_2$ savings due to use of renewable energy systems per m² of pool water surface	1,538 kg CO$_2$
Required investment per CO$_2$ emission savings (solar energy	0.7 € per kg of

and biomass)	CO_2 annual savings
Required investment per CO_2 emission savings (solar energy and GSHP)	0.96 € per kg of CO_2 annual savings
Payback period of the investment in solar energy and biomass	2.75 years
Payback period of the investment in solar energy and GSHP	3.21 years

Comparison of the abovementioned results shows that in the second case (solar energy and GSHP) the capital cost of the required energy systems is higher than in the first case. However, fuels cost in the first case (solar energy and biomass) is higher than in the second case (solar energy and GSHP) which is zero. The payback period of the two renewable energy investments based only on the annual gain due to lower fuel costs is very attractive. However, maintenance and depreciation costs should be also taken into account to get more accurate estimates regarding the economic feasibility of the investment.

6. Conclusions

Reduction of energy use and CO_2 emissions as well as higher use of renewable energies replacing fossil fuels use in buildings has been promoted by current EU policies. Creation of swimming pools with net zero CO_2 emissions due to energy use can be obtained using various renewable energy sources which could cover all their heating and power requirements. Many renewable energy technologies are currently reliable, mature and cost effective due to recent technological advances and breakthroughs. A case study in Crete, Greece has shown that creation of a net zero CO_2 emissions swimming pool is currently feasible using locally available renewable energies. These renewable energies include solar thermal energy, solar photovoltaic, solid biomass and low enthalpy geothermal energy with heat pumps. Their use decreases the carbon footprint due to operating energy use in the pool while offers economic benefits regarding the energy cost. Two combinations of renewable energy technologies can be used for covering all the heating and power needs in the swimming pool. In the first, solar thermal energy and solid biomass can be used for heat generation while solar-PV energy for electricity generation. In the second, solar thermal energy and ground source heat pumps for heat generation while solar-PV energy for electricity generation. The capital cost of the necessary renewable energy systems for zeroing its net CO_2 emissions has been estimated between 1,072 €/m² and 1,474

€/m² of pool surface. The same methodology can be used for creation of zero CO_2 emissions swimming pools in other locations than Crete with different availability of renewable energy resources. Current work contributes in the promotion of low carbon buildings in Europe according to EU directive 2009/28/EC as well as in climate change mitigation. Further work is recommended focused on the implementation of a zero CO_2 emissions swimming pool due to operating energy use in order to test, verify and assess the use of various renewable energy systems in it.

References

[1] Balat, M. & Ayar, G. (2005). "Biomass Energy in the World, Use of Biomass and Potential Trends", *Energy Sources*, 27(10), pp. 931-940.

[2] Bibbiani, C., Fantozzi, F., Gergari, C., Campiotti, C.A., Schettini, E. & Vox, G. (2016). "Wood biomass as sustainable energy for greenhouses heating in Italy", *Agriculture and Agricultural Science Procedia*, 8, pp. 637-645. DOI 10.1016/j.aaspro.2016.02.086.

[3] Carvalho, M.D.G., Bonifacio, M. & Dechamps, P. (2011). "Building a low carbon society", *Energy*, 36, pp. 1842-1847.

[4] Chiasson, A. "Residential swimming pool heating with geothermal heat pumps systems". Retrieved on 9/1/2017 from http://www.oit.edu/docs/default-source/geoheat-center-documents/publications/heat-pump/tp117.pdf?sfvrsn=

[5] Connolly, D., Lund, H., Mathiesen, B.V. & Leahy, M. (2011). "The first steps towards a 100 % renewable energy-system for Ireland", *Applied Energy*, 88, pp. 502-507.

[6] Cosic, B., Krajacic, G. & Duic, N. (2012). "A 100 % renewable energy system in the year 2050: The case of Macedonia", *Energy*, 48, pp. 80-87.

[7] Czarnecki, J.T. (1963). "A method of heating swimming pools by solar energy", *Solar Energy*, 7(1), pp. 3-7.

[8] "Energy efficiency in swimming pools – for centre managers and operators, Good practice guide 219", The department of the environment, transport and the regions' energy efficiency best practice program, Britain, 1997. Retrieved on 9/1/2017 from http://www.cibse.org/getmedia/f36a292c-8eea-4610-b764-e23774a52cb9/GPG219-Energy-Efficiency-in-Swimming-Pools.pdf.aspx

[8] EU directive 2009/28/EC on the promotion of the use of energy from renewable sources. Retrieved on 9/1/2017 from http://eur-lex.europa.eu/legal-content/EN/TXT/PDF/?uri=CELEX:32009L0028&from=EN

[9] Good, C., Andresen, I. & Hestnes, A.G. (2015). "Solar energy for net zero energy buildings - A comparison between solar thermal, PV and photovoltaic-thermal (PV/T) systems", *Solar Energy*, 122, pp. 986-996.

[10] Haraldsson, I.G. & Cordero, A.L. (2014). "Geothermal baths, swimming pools and spas: Examples from Ecuador and Iceland", Presented at *"Short Course VI on Utilization of Low- and Medium-Enthalpy Geothermal Resources and Financial Aspects of Utilization"*, organized by UNU-GTP and LaGeo, in Santa Tecla, El Salvador, March 23-29, 2014.

[11] Kampel, W. (2015). "Energy efficiency in swimming facilities", Ph.D. thesis, Norwegian University of Science and Technology, Trondheim, Norway. Retrieved on 9/1/2017 from https://brage.bibsys.no/xmlui/bitstream/handle/11250/2366793/PhD_Wolfgang_Kampel.pdf?sequence=1

[12] Liu, S., Zhang, W., Dong, Z. & Sun, G. (2015). "Analysis on several heat pump applications in large public buildings", *Journal of Building Construction and Planning Research*, 3, pp. 136-148.

[12] Lund, H. & Mathiesen, B.V. (2009). "Energy system analysis of 100 % renewable energy systems-The case of Denmark in years 2030-2050", *Energy*, 34, pp. 525-531.

[13] Mancic, M.V., Zivkovic, D.S., Milosavljevic, P.M & Todorovic, M.N. (2014). "Mathematical modeling and simulation of the thermal performance of a solar heated indoor swimming pool", *Thermal Science*, 18(3), pp. 999-1010.

[14] Mathiesen, B.V., Lund, H. & Connolly, D. (2012). "Limiting biomass consumption for heating in 100 % renewable energy systems", *Energy*, 48, pp. 160-168.

[15] Michopoulos, A., Tsikaloudaki, A., Voulgari, V. & Zachariadis, Th. (2015). "Analysis of ground source heat pumps in residential buildings", *Proceedings World Geothermal Congress 2015*, Melbourne, Australia, 19-25 April 2015.

[16] Sarachaga, L. (2008). "Solar energy use in outdoor swimming pools-SOLPOOL", EIE -06-085. Retrieved on 9/1/2017 from https://ec.europa.eu/energy/intelligent/projects/sites/iee-projects/files/projects/documents/solpool_european_report_on_the_state_of_the_demand_and_potential_of_solar_heating_of_outdoor_swimming_pools_en.pdf

[17] Scognamiglio, A. & Rostvik, H. (2012). "Photovoltaics and zero energy buildings: a new opportunity and challenge for design", *Paper presented at 27th*

EU PVSEC, Frankfurt, Germany 2012, Progress in Photovoltaics: Research and Applications. DOI: 10.1002/pip.2286

[18] Trianti-Stourna, E., Spyropoulou, K., Theofylaktos, C., Droutsa, K., Balaras, C.A., Santamouris, M., Asimakopoulos, D.N, Lazaropoulou, G. & Papanikolaou, N. (1998). "Energy conservation strategies for sport centers: Part B. Swimming pools", *Energy and Building*, 27, pp. 123-135.

[19] Tsoutsos, Th., Karagiorgas, M., Zidianakis, G., Drosou, V., Aidonis, A., Gouskos, Z. & Moeses, C. (2009). "Development of applications of solar thermal cooling systems in Greece and Cyprus", *Fresenius Environmental Bulletin*, 18(7b), pp. 1-15.

[20] Urge-Vorsatz, D., Danny Harvey, L.D., Mirasgedis, S. & Levine, M.D. (2007). "Mitigating CO_2 emissions from energy use in the world's buildings", *Building Research & Information*, 35(4), pp. 379-398. doi: 10.1080/09613210701325883

[21] Visa, I., Moldovan, M.D., Comsit, M. & Duta, A. (2014). "Improving the renewable energy mix in a building towards the nearly zero energy status", *Energy and Building*, 68, pp. 72-78.

[22] Vourdoubas, J. (2015). "Creation of hotels with zero CO_2 emissions due to energy use. A case study in Crete-Greece", *Journal of Energy and Power Sources*, 2(8), pp. 301-307.

[23] Vourdoubas, J. (2016). "Creation of zero CO_2 emissions hospitals due to energy use. A Case study in Crete-Greece", *Journal of Engineering and Architecture*, 3(2), pp. 1-9.

[10] Creation of zero CO_2 emissions office buildings due to energy use. A case study in Crete, Greece

1. Introduction

Buildings consume approximately 40 % of total energy consumption in EU. Current European policies aim to increase their energy efficiency the use of renewable energies instead of fossil fuels as well as to lower their CO_2 emissions. According to EU directive 2010/31/EU after 2018 and 2020 all new public and private buildings must be nearly zero energy buildings (NZEBs). Current research investigates the possibility of creating net zero carbon emissions office buildings due to operating energy use with reference to the island of Crete, Greece. This can be achieved with replacement of fossil fuels used for energy generation with renewable energies using mature, reliable and cost-effective technologies. Various renewable energies are currently available in Crete, Greece, including solar energy, solid biomass and low enthalpy geothermal energy. They can be used for electricity, heat and cooling generation covering all energy requirements in office buildings. Creation of net zero carbon emission buildings, including office buildings, is important for achieving EU goals for climate change mitigation as well as for carbon neutrality by 2050.

2. Literature survey

A report on cost optimal and nearly zero energy office buildings in Estonia has been presented by Pikas et al, 2014. The authors developed a methodology following three steps including: a) Selection of optimal insulation thickness, b) Development of cost efficient solutions, and c) Use of local renewable energy resources in order to achieve nearly zero energy buildings (NZEBs). The authors stated that due to cold Estonian climate existing NZEB solutions are not cost-optimal. Cost-optimal design for nearly zero energy office buildings located in warmer climates has been reported by Congedo et al, 2015. The authors studied various energy efficient measures for office buildings in Italy indicating that they could reduce energy consumption by 39% and CO_2 emissions by 41% with low cost. They mentioned that their methodology could be applied in new office buildings in warm climates in compliance with current EU policies. Data analysis on energy use for lighting in large office buildings in China has been reported by Zhou et al, 2015. The authors stated that lighting consumes about 20-40% of total electricity use in large office buildings in China. They studied 15 large office buildings in Beijing and Hong-Kong and they found that the annual average lighting energy consumption varied from 15 KWh/m^2 to 70 KWh/m^2 while the ratio of lighting energy consumption in different office buildings varied approximately by 30-70%. The influence of intelligent glazed facades on energy and comfort performance of office buildings in Denmark has been reported by Liu et al, 2015. Their results indicated that energy consumption in an office building in Denmark could be reduced by approximately 60% in the case that an ordinary façade could be replaced by an intelligent one. A comparative evaluation of optimal energy efficiency designs for French and USA office buildings has been published by Krarti et al, 2015. The authors presented a general approach for design optimization of energy efficient office buildings. They found that optimizing life cycle costs resulted on average at 30% primary energy savings in office buildings located in USA and at 40% energy savings in France. However, in order to achieve a NZEB, solar-PV panels must be placed on the roof of the building or alternative off-site if the necessary roof surface is not enough. Comparison of energy consumption in energy efficient buildings in China and in USA has been presented by Liu et al, 2014. The authors analyzed the total annual energy consumption of various energy efficient office buildings stating that the mean annual energy consumption in China was 74.83 KWh/m^2 compared to 88.81 KWh/m^2 in USA. The higher energy consumption in office buildings in USA was justified from the fact that designers pursue extraordinary indoor thermal comfort. An assessment of natural resources conservation in office buildings using TOBUS, a European methodology for office building refurbishment, has been presented by Balaras et al, 2002. The authors examined various scenarios for energy conservation in office buildings. They mentioned that energy conservation in Hellenic and Danish office buildings range for space heating from 5 % to 7 % and 0.5 % to 6 %, for space cooling from 1 % to 38 % and 4 % to 20 %, for lighting from 40 % to 53 % and 26 % to 62 %, for office equipment from 13 % to 62 % and 13 % to 87 % and for elevators at 35 % and 23 % respectively. Life-cycle energy use in office buildings has been reported by Cole et al, 1996. The authors studied the use of embodied and operating energy in an office building with covered area at 4,620 m^2 in Canada. They stated that operating energy represents the largest component of life-cycle energy use. For a typical building life of fifty (50) years the embodied energy represents only 10-20 % of its life-cycle energy use. However, the authors mentioned that if the operating energy could be reduced significantly in the future the embodied energy would represent the largest part of the life-cycle energy use in office buildings. A life-cycle energy assessment of a typical office building in Canada has been reported by Kofoworola et al, 2009. The authors studied a building with covered surface at 60,000 m² concluding that operating energy accounted for 81 % of its life-cycle energy consumption. The rest is embodied energy due to its construction at 17.4 %, to maintenance at 0.8 % and to demolition at 0.4 %. They also stated that annual energy consumption in office buildings in Thailand, Malaysia, Singapore, Indonesia and Taiwan varies between 80 KWh/m² to 300 KWh/m². A report regarding life-cycle zero energy buildings has been published by Hernandez et al, 2010. The authors defined, in accordance with the energy performance building directive, a net energy building as "a building where, as a result of a very high level of its energy efficiency, the overall primary energy consumption is equal to or less than the energy production from on-site renewable energy sources." They also stated that "net zero site energy" means that a site produces annually at least the same energy as it uses independent of the type of energy produced or used. The improvement of renewable energy mix in buildings towards nearly zero energy status has been presented by Visa et al, 2014. The authors developed a three-step methodology in order to achieve nearly zero energy status. They evaluated initially the energy status of the building, then they developed measures to reduce its energy consumption and finally they proposed on-site renewable energy generation in order to cover all its energy requirements. Use of solar energy for creating net zero energy buildings has been reported by Good et al, 2015. The authors investigated and assessed the use of three solar energy technologies in order to achieve net zero energy buildings in Norway. The three compared technologies in their study were solar thermal, solar-PV and hybrid PV-thermal. Their results indicated that high efficiency solar-PV modulus is the most appropriate technology in order to achieve zero energy balance in a Norwegian residential building. Use of solar photovoltaics in order to achieve net zero energy buildings has been reported by Scognamiglio et al, 2012. The authors stated that in zero energy buildings solar-PVs are suitable for generating electricity either on-site or off-site. They mentioned that there are challenging issues regarding the use of solar-PVs in net zero energy buildings including their integration in the building envelope. Use of solid biomass for greenhouse heating in Italy has been reported by Bibbiani et al, 2016. The authors stated that biomass boilers are cost-effective and their use for heating is highly recommended. They also reported that biomass is considered as carbon neutral fuel excluding GHG emissions during harvesting, processing and transport. Creation of net zero CO_2 emissions residential buildings in Crete has been reported by Vourdoubas, 2015. The author examined two different combinations of renewable energy sources available in Crete achieving net zero CO_2 emission buildings. The first included the use of solar thermal energy, solar-PV and solid biomass and the second the use of solar thermal energy, solar-PV and low enthalpy geothermal energy. He stated that the cost of the required renewable energy systems achieving a net zero CO_2 residential building corresponds at 10-12% of its total construction cost. Vourdoubas, 2015 has also investigated the possibility of creating net zero CO_2 emissions hospitals due to energy use implementing a case study in Crete, Greece. He examined two different scenarios estimating the investment cost of the required renewable energy systems as well as the CO_2 emission savings achieved. First the use of solar energy and solid biomass were examined and secondly the use of solar energy and ground source heat pumps. He mentioned that the investment cost in the first case was lower than in the second case. An analysis on several heat pump applications in large public buildings has been reported by Liu et al, 2015. The authors compared traditional heating systems and heat pumps used in large buildings regarding their initial investment costs. They concluded, with reference to China, that heat pumps were better in heating and cooling large buildings, regarding the economy and technology, than traditional heating systems. Creation of buildings with low energy consumption and CO_2 emissions are in the core of E.U. policies and currently a project promoting new policies for zero CO_2 emission buildings is financed by the E.U. INTERREG EUROPE program (http://www.interregeurope.eu/zeroco2/). The energy characteristics and the savings potential in office buildings in Greece have been presented by Santamouris et al, 1994. The authors monitored 186 office buildings in Greece estimating their energy consumption for heating, cooling, lighting and operation of office equipment. They mentioned that half of their energy consumption was used for heating, 10.7 % for lighting, 12.6 % for cooling and 25.9 % for operation of office equipment. The mean annual energy consumption in the 186 office buildings was estimated at 187 KWh/m^2. The authors stated that proper energy saving measures could reduce the mean annual energy consumption at 150-154 KWh/m^2. The energy consumption in office buildings in UK has been investigated (Energy consumption guide 19, 2003). Office buildings are categorized in four groups according to their air conditioning system and the type of construction. The typical energy consumption as well as good practice values in each group is mentioned. In standard office buildings which are air conditioned the typical annual energy consumption has been estimated at 226 KWh/m^2 while the gas or oil consumption at 178 KWh/m^2. For the same type of buildings the good practice annual energy consumption is at 128 KWh/m^2 while the gas or oil consumption at 97 KWh/m^2 respectively. In standard office buildings 44.1 % of total energy consumption was used in heating, 7.7 % in cooling, 13.4 % in lighting and the rest 34.8 % for operation of electric machinery. Typical energy consumption in office buildings located in Greece and in UK is presented in Table 1.

Table 1: Annual energy consumption in office buildings located in Greece and in UK

Country	Total (KWh/m^2)	Heating (KWh/m^2)	Cooling (KWh/m^2)	Lighting (KWh/m^2)	Operation of electric equipment (KWh/m^2)
Greece[1]	186	94.4	23.4	20	48.2
UK, Standard building[2]	404	178	31	54	141
UK, Best practice building[2]	225	97	14	27	87

[1]*Source: Santamouris et al, 1994,* [2]*Source: Energy consumption guide 19, 2003*

The aim of current work is to investigate the possibility of creating net zero CO_2 emissions office buildings due to energy use located in the island of Crete, Greece which is characterized by mild climate.

This can be achieved using renewable energy sources (RES) available in the island which could cover all heat and power requirements in office buildings. A methodology for creating net zero CO_2 emissions office buildings using various RES has also been developed. Finally a case study of an office building with net zero CO_2 emissions due to energy use in Crete has been presented while the characteristics of renewable energy systems used have been described.

3. Availability of renewable energy sources in the island of Crete, Greece

Renewable energy sources are abundant in the island of Crete, particularly solar energy, wind energy and solid biomass. They are currently used in power and heat generation. Solar photovoltaics and wind farms are currently used for electricity generation in the island. Solar thermal energy with simple thermoshiphonic systems is used for hot water production and solid biomass, particularly olive tree by-products and residues, is used for space heating, hot water production as well as for process heat generation in various industries. The potential of hydroelectric energy and geothermal energy is very small. However, use of ground source heat pumps for heat and cooling production in buildings is increasing. Average annual solar irradiance in Crete has been estimated at 1,700-1,880 KWh/m^2 (Kagarakis, 1987). Solar thermal systems are extensively used for hot water production in buildings during the last 30 years and this technology is reliable, mature and cost-effective. Solar-PVs use is growing partly due to the fact that there has been a sharp decrease in their prices during the last few years. Solar panels are placed on building roofs generating electricity either with feed-in tariffs initiative or with the net-metering regulations while their use is profitable. Solid biomass is broadly used in Crete for many years for heat generation in buildings, in agriculture and in industry. Due to many olive tree orchards in the island various olive tree byproducts and residues are produced annually. It has been estimated that annual olive kernel wood production in Crete is approximately 110,000 tons (Vourdoubas, 2015). Since solid biomass is a rather cheap heating fuel compared with other heating fuels and electricity its consumption has rapidly increased partly due to the current economic crisis in Greece. Low enthalpy geothermal energy with heat pumps is increasingly used for heat and cooling production. However, their high installation cost is counterbalanced in the long term by their high efficiency. Wind energy is extensively used for power generation in large wind farms while use of small wind turbines in buildings is rather limited. Use of renewable energy systems for energy generation in buildings in Crete is presented in table 2.

Table 2: Use of various renewable energies in buildings in Crete

Energy source/technology	Energy produced	Main use	Availability
Solar thermal	heat	Domestic hot water production	High
Solar-PV	electricity	Electricity generation	High
Solid biomass	heat	Space heating, hot water production	High
Low enthalpy geothermal energy with heat pumps	heat and cooling	Heat and cooling, , hot water production	High

Source: Own estimations

4. Creation of zero CO₂ emissions office buildings

Creation of zero CO_2 emissions office buildings due to operating energy use can be achieved if the following conditions are fulfilled:

a) Only renewable energy sources are used for covering all their heating needs, and

b) Annual grid electricity used in an office building should be offset with solar-PV electricity generated in the building. Photovoltaic panels can be installed in the building roof; if there is not enough space, they could be located off-site.

Energy consumption in the building could be reduced with various energy-saving measures and most of them are cost-effective. In this case the size of the necessary renewable energy systems providing all the heat and electricity required will be lower reducing their installation cost as well. Although reduction of energy use in buildings is desired the target of net zero CO_2 emissions can be obtained without lowering their initial energy consumption. When there is no fossil fuel consumption in an office building and the grid electricity used, which is mainly derived from fossil fuels, has been replaced by solar green electricity its carbon emissions will be zero. The embodied energy in the building which in a life span of fifty (50) years corresponds approximately at 10-20 % of its total life cycle energy has not been taken into account. Neither the embodied energy of renewable energy systems used in the office building has been included. It has also been assumed that solid biomass, used for heat generation, has zero CO_2 emissions. Therefore, CO_2 emissions due to biomass harvesting, processing and transportation have not been taken into account.

5. Design of an office building with zero CO₂ emissions in Crete, Greece

Renewable energy sources available in Crete could be used for covering all energy requirements in buildings located in Crete, Greece. Solid biomass or ground source heat pumps could be used for heat and cooling production combined with solar-PVs for electricity generation. Estimation of the necessary renewable energy systems in an office building located in Crete with covered surface at 1,000 m² is followed. It is assumed that its energy consumption will be similar with the energy consumption in Greek buildings reported by Santamouris et al, 1994. Therefore, its total annual energy consumption is 186,000 KWh/m² distributed at 94,400 KWh/m² for heating, 23,400 KWh/m² for cooling, 20,000 KWh/m² for lighting and 48,200 KWh/m² for operation of various electric devices. It is also assumed that the office building is connected with the electric grid and all its annual electricity consumption is offset by solar-PV electricity according to net-metering regulations. However, since part of grid electricity has been generated with renewable energies the solar-PV system is oversized.

5.1 Use of solar energy and solid biomass for covering all its energy requirements

Solar-PV energy and solid biomass can be used for covering the electricity and heating needs in the building. Total annual electricity and heating needs in the abovementioned office building are estimated at 91,600 KWh and 94,400 KWh respectively. It is assumed that woody biomass will be used with heat content at 4,200 Kcal/kg while the efficiency of the biomass burning system is 75 %. The price of solid biomass in Crete is at 150 €/ton. The biomass boiler will operate 1,200 hours per year and its power will be 78.7 KW$_{th}$. Annual consumption of solid biomass will be at 26.3 tons and its cost will be at 3,945 €. It has also been assumed that electricity generation from solar-PV panels in Crete is 1,500 KWh/KW$_p$ while their unit cost is 1,200 €/KW$_p$. The power of the solar-PV system providing all electricity required in the office building is 61.1 KW$_p$ while its cost is 73,320 €. In order to estimate CO_2 emissions savings due to renewable energies use in the office building it has been assumed that the CO_2 emissions coefficient for heating oil is 0.32 kg CO_2/KWh while for grid electricity 0.75 kg CO_2/KWh. If only heating oil and grid electricity were used in the office building then its CO_2 emissions will be at 68.7 tons CO_2/year due to electricity use and 30.2 tons CO_2/year due to heating oil use, totally 98.9 tons CO_2/year. The design characteristics and the cost of the renewable energy systems as well as the CO_2 emission savings are presented in Table 3.

Table 3: Characteristics of solid biomass burning and solar-PV systems providing all required energy in an office building in Crete

Annual heat consumption in the building	94,400 KWh
Annual electricity consumption in the building	91,600 KWh
Power of the biomass burning system	78.7 KW$_{th}$
Cost of the biomass burning system (250 €/KW$_{th}$)	19,675 €
Annual consumption of biomass	26.3 tons
Cost of the required biomass	3,945 €/year
CO_2 emissions savings due to biomass use [1]	30.2 tonsCO_2/year
Nominal power of the solar-PV system	61.1 KW$_p$
Cost of the solar-PV system	73,320 €
CO_2 emissions savings due to generation of solar-	68.7 tonsCO_2/year

electricity	
Total capital cost of renewable energy systems	92,995 €
Total CO₂ emissions savings	98.9 tonsCO₂/year
Total annual operating cost due to use of renewable fuels	3,945 €/year

[1] It is assumed that CO_2 emissions due to biomass use are zero.

For a better estimation of the total annual operating cost of the abovementioned renewable energy systems maintenance and depreciation costs must be also included.

5.2 Use of solar energy and low enthalpy geothermal energy for covering all its energy requirements

The overall energy needs of the office building could be covered with a ground source heat pump (GSHP) providing heating and cooling and a solar-PV system generating annually all the electricity required. The heating and cooling needs are 117,800 KWh annually. Assuming that the C.O.P. of the GSHP is 4.5 it will consume 26,178 KWh per year during its operation. Since the annual needs for lighting are 20,000 KWh and for operation of various equipment 48,200 KWh the total electricity needs in the building are 94,378 KWh. Assuming that the heat pump will operate 2,200 hours per year its estimated power will be 12 KW_{el} (54 KW_{th}). Design characteristics of the renewable energy systems as well as their capital cost and CO_2 emissions savings achieved are presented in Table 4.

Table 4: Various characteristics of the renewable energy systems generating all the required energy in the office building in Crete

Annual heat and cooling needs in the building	117,800 KWh
C.O.P. of the heat pump	4.5
Operating hours of the heat pump	2,200 hours/year
Electricity consumption of the heat pump	26,178 KWh/year
Total electricity consumption in the building	94,378 KWh/year
Power of the heat pump	12 KW_{el}
Cost of the GSHP (1,800 €/KW_{el})	21,600 €
Nominal power of the solar-PV system	62.9 KW_p
Cost of the solar-PV system	75,480 €
CO_2 emissions savings due to use of renewable energies	98.9 tonsCO₂/year

Total capital cost of renewable energy systems	97,080 €

A comparison of the different renewable energy systems installed in the office building in Crete in order to zero its net CO$_2$ emissions is presented in Table 5.

Table 5: Comparison of different renewable energy systems installed in the office building in Crete for zeroing its net CO$_2$ emissions

Parameter	Solid biomass and solar-PV	GSHP and solar-PV
Capital cost	92,995 €	97,080 €
Annual operating cost due to fuel use	3,945 €/year	0
CO$_2$ emission savings	98.9 tonsCO$_2$/year	98.9 tonsCO$_2$/year

6. Discussion and Conclusions

The work presented is original investigating the possibility of creating net zero CO$_2$ emissions office buildings due to operating energy use. Current EU policies promote the creation of nearly zero energy buildings having nearly zero CO$_2$ emissions. Depending on the availability of renewable energy sources in a location the target of zero CO$_2$ emissions buildings can be achieved using a combination of benign green energy sources. With reference to the island of Crete, Greece it has been indicated that using solar-PV energy, solid biomass and low enthalpy geothermal energy the target of net zero CO$_2$ emissions buildings could be achieved. In order to obtain this goal fossil fuels are not used and grid electricity used in the office building is offset by green solar electricity according to net-metering regulations. The abovementioned renewable energies are abundant in Crete while their technologies are mature, reliable and cost-effective used already in various applications for heat and power generation. Two different combinations of locally available renewable energy sources can be used in order to achieve net zero CO$_2$ emissions office buildings in Crete. The first includes use of solid biomass and solar-PV energy and the second use of ground source heat pumps and solar-PV energy. Cost analysis indicated that the cost of the necessary investments in renewable energy systems in office buildings in Crete zeroing their net CO$_2$ emissions are slightly less than 100 €/m^2. In the previous analysis the embodied energy in the office building has not been taken into account neither the embodied energy of the renewable energy systems used for energy generation. Therefore, more accurate results can be achieved if a life cycle analysis is implemented. The results of this study could be used for the creation of net zero CO_2 emissions office buildings due to operating energy use in various locations probably with different availability of renewable energies. They are also complying with current EU policies for transforming the continent in a low carbon economy as well as for reducing emissions of GHGs and mitigating climate change.

References

[1] EU directive 2010/31/EU on the energy performance of buildings. Retrieved on 31/1/2017 from http://eur-lex.europa.eu/LexUriServ/LexUriServ.do?uri=OJ:L:2010:153:0013:0035:en:PDF

[2] Pikas, E., Thalfeldt, M. & Kurnitski, J. (2014). "Cost optimal and nearly zero energy building solutions for office buildings", *Energy and Buildings*, 74, pp. 30-42.

[3] Congedo, P.M., Baglivo, C., D'Agostino, D. & Zaca, I. (2015). "Cost-optimal design for nearly zero office buildings located in warm climates", *Energy*, 91, pp. 967-982.

[4] Zhou, X., Yan, D., Hong, T. & Ren, X. (2015). "Data analysis and stochastic modeling of lighting energy use in large office buildings in China", *Energy and Buildings*, 86, pp. 275-287.

[5] Liu, M., Wittchen, K.B. & Heiselberg, P.K. (2015). "Control strategies for intelligent glazed façade and their influence on energy and comfort performance of office buildings in Denmark", *Applied Energy*, 145, pp. 43-51.

[6] Krarti, M. & Deneuville, A. (2015). "Comparative evaluation of optimal energy efficiency designs for French and US office buildings", *Energy and Buildings*, 93, pp. 332-344.

[7] Liu, L., Zhao, J., Liu, X. & Wang, Z. (2014). "Energy consumption comparison analysis of high energy efficiency office buildings in typical climate zones of China and U.S. based on correction model", *Energy*, 65, pp. 221-232.

[8] Balaras, C.A., Droutsa, K., Argiriou, A.A. & Wittchen, K. (2002). "Assessment of energy and natural resources conservation in office buildings using TOBUS", *Energy and Buildings*, 34, pp. 135-153.

[9] Cole, R.J. & Kernan, P.C. (1996). "Life cycle energy use in office buildings", *Building and Environment*, 31(4), pp. 307-317.

[10] Kofoworola, O.F. & Gheewala, S.H. (2009). "Life cycle energy assessment of a typical office building in Thailand", *Energy and Buildings*, 41, pp. 1076-1083.

[11] Hernandez, P. & Kenny, P. (2010). "From net energy to zero energy buildings: Defining life cycle zero energy buildings (LC-ZEB)", *Energy and Buildings*, 42, pp. 815-821.

[12] Visa, I., Moldovan, M.D., Comsit, M. & Duta, A. (2014). "Improving the renewable energy mix in a building towards the nearly zero energy status", *Energy and Buildings*, 68, pp. 72-78.

[13] Good, C., Andresen, I. & Hestmes, A.G. (2015). "Solar energy for zero energy buildings – A comparison between solar thermal, PV and photovoltaic thermal (PV/T) systems", *Solar Energy*, 122, pp. 986-996.

[14] Scognamiglio, A. & Rostvik, H.N. (2012). "Photovoltaics and zero energy buildings: A new opportunity and challenge for design", *in Progress photovoltaics: Res Appl.*, at 27th EU PVSEC, Frankfurt, Germany.

[15] Bibbiani, C., Fantozzi, F., Gargari, C., Campiotti, A., Schettini, E. & Vox, G. (2016)."Wood biomass as sustainable energy for greenhouse heating in Italy", *Agriculture and Agricultural Science Procedia*, 8, pp. 637-645.

[16] Vourdoubas, J. (2015). "Creation of Zero CO_2 Emissions Residential Buildings Due to Energy Use: A Case Study in Crete-Greece", *Journal of Civil Engineering and Architecture Research*, 3, pp. 1251-1259.

[17] Vourdoubas, J. (2015). "Creation of Zero CO_2 Emissions Hospitals Due to Energy Use. A Case Study in Crete-Greece", *Journal of Engineering and Architecture*, 3, pp. 1-9.

[18] Liu, S., Zhang, W., Dong, Z. & Sun, G. (2015). "Analysis on several heat pump applications in large public buildings", *Journal of Building Construction and Planning Research*, 3, pp. 136-148.

[19] "Promotion of near zero CO_2 emission buildings due to energy use", Interreg Europe project. Retrieved on 31/1/2017 from http://www.interregeurope.eu/zeroco2/

[20] Santamouris, M., Argiriou, A., Dascalaki, E., Balaras, C. & Gaglia, A. (1994). "Energy characteristics and saving potential in office buildings", *Solar Energy*, 52(1), pp. 59-66.

[21] "Energy use in offices, energy consumption guide 19", Best practice program. Retrieved at 31/1/2017 from http://www.cibse.org/getmedia/7fb5616f-1ed7-4854-bf72-2dae1d8bde62/ECG19-Energy-Use-in-Offices-(formerly-ECON19).pdf.aspx

[22] Kagarakis, K. "Photovoltaic Technology" *Symmetria Publications*, Athens, 1987, (In Greek).

[23] Vourdoubas, J. (2015). "Present and future uses of biomass for energy generation in the island of Crete-Greece", *Journal of Energy and Power Sources*, 2, pp. 158-163.

[11] Energy consumption and carbon emissions in Venizelio hospital in Crete, Greece. Can it become carbon neutral?

1. Introduction

European policies currently promote the transition of the continent to a low-carbon economy in order to mitigate climate change which consists of the major global environmental threat. For achieving this target energy efficiency must be improved while renewable energies should replace fossil fuels in all sectors of economy. Hospital buildings consume large amounts of energy compared with residential and commercial buildings due to their continuous and complex operation. Their potential in reducing energy consumption and carbon emissions is high. Energy consumption in hospital buildings varies depending on the type and year of buildings construction, the mode of operation, their location and the local climate. They mainly consume conventional fuels while the use of benign energy sources for heat and electricity generation is rather limited. Current work is focused in the investigation of the possibility of achieving net zero carbon emission hospitals due to energy use with reference Venizelio hospital located in Crete, Greece. Creation of carbon neutral hospitals is important for achieving the national and European targets for lower GHG emissions and climate change mitigation.

2. Literature survey

2.1 Energy consumption in hospitals

Hu et al, 2004 have reported on energy consumption and energy cost in a large acute hospital in Taiwan. The authors stated that air conditioning has a share at 52% in total energy consumption, lighting at 12% and other utilities at 36%. They also mentioned that annual energy use in the hospital was at 259.45 KWh/m^2 while 88% of its annual heat requirements were covered with electricity and the remaining 12% with gas. Moghimi et al, 2011 have reported on energy consumption in a large scale educational hospital in Malaysia. The authors stated that its annual energy consumption was at 245 KWh/m^2 while electricity had a share at 75% on that. A best-practice program for energy consumption in hospitals has been published by the Department of Environment, Transport and the region's energy efficiency, UK, 1999. The report stated that currently the annual heat energy consumption in a typical hospital in UK is 445 KWh/m^2 while its annual electricity consumption is 160 KWh/m^2. A good practice target for the same hospital is at 339 KWh/m² for annual heat consumption and at 103 KWh/m² for annual electricity consumption. Energy consumption in Brazilian hospitals has been reported by Szklo et al, 2004. The authors stated that annual energy consumption for a medium-size hospital was at 230-460 KWh/m² while more electricity than heating fuel was used. Santamouris et al, 1994 have reported on energy performance and energy conservation in health care buildings in Greece. The authors studied 30 healthcare buildings and they stated that their annual energy consumption was at 275-407 KWh/m². They also mentioned that space heating required 65.3-73.4 % of the total energy consumption while use of energy saving techniques could reduce the overall energy use by 10%. Kolokotsa et al, 2012 have reported on energy conservation techniques in hospital buildings. The authors reported that use of state-of-the-art energy-saving technologies could reduce energy consumption by 10%. They also mentioned that energy breakdown in a good practice 500-bed hospital is 34% for space heating, 14% for base load, 14% for lighting and 38% for other electricity uses. Santamouris et al, 1994 have reported on the use of renewable energies and energy conservation technologies in various buildings in Europe. The authors stated that annual energy consumption per sector in a typical Greek hospital is 3 KWh/m² for cooling, 299 KWh/m² for heating, 52 KWh/m² for lighting, and 53 KWh/m² for the operation of various equipment, in total 407 KWh/m². Sofronis et al, 2000 have reported on energy consumption in Greek hospitals. The authors stated that annual energy consumption varies according to climate zone from 270 KWh/m² in south, to 438 KWh/m² in north. The average annual energy consumption was at 370 KWh/m² distributed at 290 KWh/m² for heating and 80 KWh/m² for electricity. A report on energy saving in hospitals has been published by CADDET, 1997. According to this report the annual energy consumption in hospitals varies between 250-1,000 KWh/m² while use of appropriate energy-saving techniques could result in energy reduction by 20-44 %. The report mentioned that annual energy consumption in Greek hospitals is estimated at 300 KWh/m² for heating and 110 KWh/m² for electricity. Van Schijndel, 2002 has reported on co-generation of heat and power in hospitals for covering their heating, cooling and electricity needs while the technology is profitable. Bujak, 2010 has reported on heat consumption in hospitals. The author stated that in Poland, during 2003-2008, the annual energy consumption in large hospitals with over 600 beds varied between 250-333 KWh/m². Short et al, 2009 have studied the use of low energy ventilation and cooling strategies in the design of new hospitals. The authors mentioned that they will decrease energy consumption and they will be profitable. Saidur et al, 2010 studied the energy use in Malaysian hospitals indicating that use of efficient electric motors could significantly decrease energy consumption while they can be highly profitable. Herrera et al, 2003 have studied the use of pinch technology in hospitals reporting that power savings at 38% could be achieved. Argiriou et al, 1994 have investigated the energy consumption and the indoor air quality in office and hospital buildings located in Athens, Greece. They reported that concentration of NO_x and SO_2 did not exceed the upper limits of the World Health Organization (WHO).

2.2 Use of renewable energies in hospitals

There are not many reports published so far regarding the use of renewable energy technologies in hospitals. During the implementation of an Intelligent Energy Europe project, RES-hospitals, the use of renewable energies in European hospitals was studied. Paksoy et al, 2000 have designed a solar heating system combined with heat storage in an underground aquifer for providing heat and cooling in Balcali hospital located in Adana, Turkey. The authors stated that their simulation results indicated that an underground aquifer with dimensions 350X400 meters would be enough for storage of roughly 14,000 MWh/year in the form of heat and cold at +10°C. Buonomano et al, 2014 have reported a novel renewable poly-generation system for a hospital building located in Naples, Italy. The system was integrating concentrated photovoltaic thermal collectors combined with solar heating and cooling technologies. The authors stated that the energy performance of the system was excellent since the generated energy was consumed in-situ while its payback period was approximately 12 years. Vourdoubas, 2015 has reported on the creation of zero CO_2 emission hospitals due to operating energy use with reference the island of Crete, Greece. The share of electricity in total energy consumption was at 42% while for heating at 58%. He also estimated that the investment cost of the necessary renewable energy technologies covering all the energy needs in hospitals in Crete, Greece varies between 7,434 €/bed to 8,679 €/bed. Vourdoubas, 2016 has reported on using solid biomass for covering all the heating needs in hospitals in Crete, Greece. He has proposed the use of olive kernel wood, a by-product of the olive processing industry, for covering all the annual heating needs in existing hospitals which have been assumed at 300 KWh/m^2. The author estimated that the investment cost of the required biomass heating systems varies between 1,963 €/bed to 2,328 €/bed. Use of geothermal energy in a hospital in New Zealand has been reported by Steins et al, 2012. The heating requirements in the hospital were covered by high enthalpy steam which can be also used in the future for electricity generation. A case study concerning the use of a solar absorption cooling system in a Greek hospital has been reported by Tsoutsos et al, 2010. The authors indicated that its use is cost-effective. An aquifer thermal storage system in a Belgian hospital has been reported by Vanhoudt et al, 2011. The thermal energy storage system was combined with a heat pump for space heating and cooling in the buildings. The authors stated that the primary energy consumption of the heat pump was 71% lower than the conventional energy system using gas-fired boilers. They also mentioned that the payback period of the green energy investment was 8.4 years without any subsidies. Bizzari et al, 2006 have reported on the use of new technologies for effective retrofitting in hospitals. The authors studied the use of a phosphoric acid fuel cell, a solar thermal system and a solar photovoltaic system, concluding that their use in hospital's refurbishment could be profitable. Bizzari et al, 2004 have also studied the use of a fuel cell hybrid plant in a hospital in northern Italy, mentioning that its use could result in significant heat and power savings. Renedo et al, 2006 have studied four different co-generation alternatives for providing heat, air-conditioning and hot water in a large hospital in Spain stating that all of them could be profitable. They also suggested that tri-generation technology should be promoted in southern Europe. Ziher et al, 2006 have studied the economics of a tri-generation system with natural gas turbines for power, heat and cooling generation in a hospital in Slovenia. The authors mentioned that the system could be profitable having a low pay-back period and a high present net value.

The aims of the current study are:

a) The estimation of energy consumption and carbon emissions in Venizelio hospital located in Crete, Greece,

b) The investigation of the possibility of using various locally available renewable energies for covering the energy needs of the hospital, reducing or zeroing its carbon emissions due to energy use, and

c) The cost estimation of renewable energy investments as well as the fuel cost savings when renewable energies are used.

3. Description of the hospital

Venizelio hospital located in Heraklion, Crete, Greece was established in 1967 and since then it was gradually expanded. Its covered area is 26,172 m^2, its capacity is 440 beds and its staff includes 1,000 employees. Most of its energy systems are old with low energy efficiency. Their replacement with new and more efficient energy systems is going to reduce its overall energy consumption and it could be cost-effective. The hospital covers all its energy needs with grid electricity and heating oil. Unfortunately, renewable energy sources are not currently used in it. The energy characteristics of the hospital and its energy consumption are presented in Table 1.

Table 1. Energy consumption and carbon emissions in Venizelio hospital located in Heraklio, Crete, Greece (2015)

Total heating power [1]	6,105 MW
Total cooling power [1]	2,455 MW
Total electric power [1]	1,400 KVA
Number of beds	440
Annual electricity consumption [2]	4,895 MWh$_{el}$
Annual heating oil consumption [2]	2,444 KWh$_{th}$
Annual total energy consumption [2]	7,339 MWh
Annual total energy consumption per bed [2]	16.7 MWh/bed
Annual cost of electricity	680,000 €/year
Annual cost of heating oil	263,000 €/year
Total energy cost	943,000 €/year
Total annual energy cost per bed	2,143 €/bed
Specific annual energy consumption	280.4 KWh/m^2
CO_2 emissions due to electricity use [3]	3,671 tonsCO_2/year
CO_2 emissions due to heating oil use [3]	733 tonsCO_2/year
Total CO_2 emissions due to energy use [3]	4,404 tonsCO_2/year
Annual total CO_2 emissions due to energy use per bed [3]	10 tonsCO_2/bed
Specific annual CO_2 emissions [3]	168 kgCO_2/m^2

[1,2]*Personal communication,* [3]*Carbon emission coefficients: electricity, 0.75 kgCO$_2$/KWh, heating oil, 0.30 kgCO$_2$/KWh*

Energy is used in the hospital for space heating and cooling, hot water and steam production, lighting and operation of various electric appliances and devices. Both electricity and heating energy are used in order to cover all its energy requirements. It is assumed that energy distribution and consumption in various sectors of the hospital follow the pattern presented in table 2 according to data published in various studies.

Table 2. Distribution of energy consumption in Venizelio hospital[1]

Sector	Energy consumed (%)	Annual energy consumption (MWh)
Space heating	34	2,495
Hot water production	13	954
Space cooling	18	1,321
Lighting	14	1,028
Other electrical uses	21	1,541
Total	100	7,339

[1] *Santamouris et al, 1994; Sofronis et al, 2000; Kolokotsa et al, 2012; Hu et al, 2004*

4. Energy use and carbon emissions in hospitals

Fossil fuels and electricity derived by them dominate in the energy mix in most hospitals. Use of renewable energies in them is rather limited so far. However, current advances in renewable energy technologies combined with appropriate governmental policies for their promotion are expected to change the landscape increasing their use in hospitals. Current advances in energy-saving technologies can also reduce their energy consumption and carbon emissions. Depending on hospital's location there are various renewable energies which are mature, reliable and cost-effective which can be used for generation of electricity, heat and cooling. In the case of Crete, Greece they include solar energy, solid biomass and low enthalpy geothermal energy. Carbon emissions in hospitals depend on the share of fossil fuels in their energy mix. In order to zero CO_2 emissions due to energy use in them the following criteria must be fulfilled:

a) Fossil fuels must be replaced by renewable energies in heat generation and if possible in power generation,

b) The use of high-efficiency energy technologies like co-generation systems and heat pumps must be encouraged, and

c) The same amount of grid electricity used annually must be offset by green electricity derived by renewable energies, preferably solar-PV electricity. Current regulations in various countries including net-metering allow that.

5. Use of various renewable energy technologies for covering the energy requirements in Venizelio hospital

Various renewable energy sources locally available can be used for covering the energy requirements in Venizelio hospital. They include solar thermal energy, solar-PV, solid biomass and low enthalpy geothermal energy with high efficiency heat pumps. The high solar irradiance in Crete allows the use of solar energy for heat and electricity generation. Availability of large quantities of olive kernel wood in the island allows its use for heat production. Their technologies are mature, well-proven, reliable and cost-effective.

5.1 Use of solar thermal energy

Solar thermal energy can be used for hot water production in the hospital. Flat-plate solar collectors can be placed on the roof terrace of the buildings while the hot water would be stored in well-insulated water storage tanks. The annual heat energy of the required hot water has been estimated at 954 MWh. Assuming that the annual heat generation from flat-plate solar thermal collectors in Crete is 700 KWh/m^2 their necessary surface in order to generate all the hot water required in Venizelio hospital is estimated at 1,363 m^2.

5.2 Use of solid biomass

Solid biomass can be used for providing all the energy required for space heating in the hospital. Various types of locally produced biomass in Crete can be used including olive tree wood and olive kernel wood. Olive kernel wood has very good burning characteristics, its price is low compared with its heating value while it is used broadly for space heating in Crete. Its average annual production in the island is 110,000 tons and it is currently consumed for heating in residential and commercial buildings, greenhouses as well as in industry. It is assumed that the annual requirements for space heating in Venizelio hospital are 2,495 MWh the heating value of olive kernel wood is 3,800 kcal/kg and the efficiency of the heating system is 75%. The necessary quantity of olive kernel wood in order to produce annually all the heat required for space heating in the hospital is estimated at 754 tons. When solar heaters are not used, olive kernel wood can be alternatively used for hot water production. The thermal power of the biomass burning system is estimated at 854 KW$_{th}$.

5.3 Use of low enthalpy geothermal energy with heat pumps

Ground source heat pumps can be used for covering the heating and cooling needs in Venizelio hospital. These are energy efficient devices using electricity having COP in the range of 3-4. Assuming that the annual needs for space heating in Venizelio hospital are 2,495 MWh and for space cooling 1,321 MWh the required annual electricity consumption of the heat pumps can be estimated. Assuming that the COP of the ground source heat pumps used is 3.5 the annual required electricity for their operation would be 1,090 MWh. The required power of the heat pump for covering the peak heating and cooling loads in Venizelio hospital has been estimated at 1 MW$_{el}$. The electricity used by the ground source heat pumps can be generated by solar-PV systems. Therefore, the heating and cooling needs in the hospital can be covered with benign, green, carbon neutral energy sources.

5.4 Use of solar-PV energy

Solar-PV systems are currently used in Crete for power generation due to high solar irradiance in the island. Solar-PV energy can be used for electricity generation in the hospital. Generated electricity could be injected into the grid either with attractive feed-in tariffs or with net-metering regulations. It is estimated that solar-PV panels without tracking systems in Crete annually generate 1,500 KWh/KW$_p$. Therefore, when solid biomass is used for space heating, the nominal power of solar-PV panels generating all the required electricity in Venizelio hospital is 2,593 KW$_p$. In the case that low enthalpy geothermal heat pumps are used for space heating instead of solid biomass, electricity requirements will be higher. In this case total annual electricity requirements will be higher than in previous case at 4,980 MWh and the size of the solar-PV system will be higher at 3,320 KW$_p$. The size of various renewable energy systems which can cover all energy requirements in Venizelio hospital are presented in Table 3.

Table 3. Size of various renewable energy systems used in Venizelio hospital covering all its energy requirements

Renewable energy system	Energy generated	Annual energy production (MWh)	Size
Solar thermal system	Hot water	954	Area of flat plate collectors, 1,363 m^2
Solar-PV system	Electricity (in the	3,890	2,593 KW$_p$

	case that biomass is used for space heating)		
Solar-PV system	Electricity (in the case that ground source heat pumps are used for space heating)	4,980	3,320 KW$_p$
Solid biomass burning system	Heat for space heating	2,495	854 KW$_{th}$
Ground source heat pumps	Energy for space heating and cooling	3,816	1,000 KW

6. Cost estimations

The investment cost of the necessary renewable energy systems generating all the energy required in Venizelio hospital has been estimated knowing their size and their unit costs. It has been assumed that unit costs of renewable energy systems are: a) For the solar thermal system, 700 € per m² of flat-plate collectors, b) For the solar-PV system 1,200 €/KW$_p$, c) For the solid biomass burning system 300 €/KW$_{th}$ and d) For the ground source heat pump 2,000 €/KW. The investment costs of renewable energy systems generating all the energy used in the hospital are presented in Table 4 for the two scenarios examined. In the first scenario it is assumed that solar-thermal energy is used for hot water production, solid biomass for space heating and a solar-PV system for electricity generation. In the second scenario a solar thermal system is used for hot water production, a ground source heat pump for space heating and cooling and a solar-PV system for electricity generation.

Table 4. Investment costs of various renewable energy systems generating all the energy required in Venizelio hospital

Renewable energy system	Unit cost of the energy system	Cost: First scenario	Cost: Second scenario
Solar thermal system	700 €/m²	954.100 €	954.100 €
Solar-PV system	1,200 €/KW$_p$	4,688,000 €	5,976,000 €

Solid biomass burning system	300 €/KW$_{th}$	256,200 €	-
Low enthalpy geothermal heat pumps	2,000 €/KW	-	2,000,000 €
Total investment cost		5,898,300 €	8,930,100 €
Total cost per bed		12,405 €/bed	20,296 €/bed
Total investment cost per KWh consumed annually in the hospital		0.80 €/KWh	1.22 €/KWh
Total investment cost per annual CO$_2$ emissions savings due to energy use		1.34 €/kgCO$_2$	2.03 €/kgCO$_2$
Annual savings in fuel cost		867,600 €	943,000 €
Annual savings in fuel cost per bed		1,972 €/bed	2,143 €/bed

Table 4 indicates that in the case of using a ground source heat pump the investment cost is significantly higher than in the second case of using a solid biomass heating system due to high installation cost of heat pumps. Use of renewable energy systems for energy generation in Venizelio hospital reduces the annual fuel costs. In the first scenario examined fuel costs are related with the cost of solid biomass while in the second scenario fuels costs are zero. Assuming that the cost of olive kernel wood in Crete is 0.1 €/kg the annual consumption of 754 tons of olive kernel wood in the hospital will cost 75,400 €.

Taking into account that currently the annual total energy cost in Venizelio hospital is 943,000 € it is concluded that use of renewable energy technologies providing all the energy needed will result in significant energy cost savings.

7. Discussion

Hospital buildings consume large amounts of energy due to their complex structure compared with other residential, commercial and industrial buildings. Heat and cooling in hospitals have a high share in total energy consumption while grid electricity and heating oil are currently used. Use of well known and cost effective energy saving techniques could reduce their total energy consumption by at least 10%. Use of renewable energies in Greek hospitals is rather limited. However, due to current European policies their use is expected to increase in the future. Additionally, recent technological innovations have resulted in improvements in their reliability and cost effectiveness promoting their use in many sectors. Various renewable energies are available in Mediterranean region like solar energy, solid biomass and low enthalpy geothermal energy. All of them have been used to some extent in various hospitals with satisfactory results. They are also broadly used in residential and commercial buildings. Additionally, electricity and heat energy requirements in hospitals throughout the year favor the use of energy efficient co-generation or tri-generation systems in them. Unfortunately, natural gas is not currently available in Crete but LPG can replace it for fuelling these energy systems. Due to deep economic crisis in Greece there is a lack of private financial resources funding the necessary green energy investments in hospitals. Cooperation with energy-saving companies through energy-performance contracting can assist in funding the necessary benign energy investments. Apart from operating energy hospitals consume energy during their construction, refurbishment and demolition which is considered as their embodied energy. In order to zero their life cycle carbon emissions both operating and embodied energy should be taken into account.

8. Conclusions

Annual energy consumption in Venizelio hospital in Crete, Greece is estimated at 280.4 KWh/m^2 which is lower than the values reported for other Greek and European hospitals. Its annual carbon emissions due to operating energy use are estimated at 168 kg CO_2/m^2. Electricity share in the energy mix is approximately double compared with that of heating oil. However renewable energies for heat and power generation in the hospital are not used so far. Various renewable energies which are abundant in Crete could be used for heat and power generation. They include solar thermal energy, solar-PV energy, solid biomass and low enthalpy geothermal energy with heat pumps. These benign energy sources are already used in Crete in various applications and their technologies are mature, reliable and cost-effective. Their use could reduce or zero carbon emissions in hospital buildings due to energy use reducing or even zeroing their annual cost of fossil fuels currently used. It has been estimated that combined use of solar thermal energy, solid biomass and solar-PV energy for covering all the energy needs will require an investment capital at 5,898,300 € resulting in annual fuel cost saving at 867,600 €. Alternatively, the combined use of solar thermal energy, solar-PV energy and ground source heat pumps will require an investment capital at 8,930,100 € resulting in annual fuel cost saving at 943,000 €. Cost analysis indicated that replacement of fossil fuels with renewable energies in Venizelio hospital would be profitable resulting in economic benefits as well as environmental and social benefits. Further work should investigate the possibility of using natural gas or LPG for co-generation or tri-generation of heat, power and cooling in Venizelio hospital. Additionally, offsetting of the embodied energy in hospital buildings with green energy should be studied and the size of the necessary green energy systems should be estimated.

References

[1] Hu, S.C., Chen, J.D. & Chuah, Y.K. (2004). "Energy cost and consumption in a large acute hospital", *International Journal of Architectural Science*, 5(1), pp. 11-19.

[2] Moghimi, S., Mat, S., Lim, C.H., Zaharim, A. & Sopian, K. (2011). "Building energy index (BEI) in large scale hospitals: Case study of Malaysia", *Recent Advances in Geography, Geology, Energy, Environment and Biomedicine,* pp. 167-170. 4th WSEAS International Conference on Engineering Mechanics, Structures, Engineering Geology, EMESEG'11, 2nd International Conference on Geography and Geology 2011, WORLD-GEO'11, 5th International Conference on EDEB'11 - Corfu Island, Duration: 14 Jul 2011 → 16 Jul 2011

[3] "Energy consumption in hospitals, Energy consumption guide 72", (1999). The Department of the Environment, Transport and the Region's Energy Efficiency Best Practice program. Retrieved on 2/5/2018 from https://www.cibse.org/getmedia/a9ab0fc1-97ed-4048-b6b5-936116334bc4/ECG72-Energy-Consumption-in-Hospitals-1999.pdf.aspx

[4] Szklo, A.S., Soares, J.B. & Tolmasquim, M.T. (2004). "Energy consumption indicators and CHP potential in the Brazilian hospital sector", *Energy Conversion and Management*, 45(13-14), pp. 2075-2091. doi:10.1016/j.enconman.2003.10.019

[5] Santamouris, M., Dascalaki, E., Balaras, C., Argiriou, A. & Gaglia, A. (1994). "Energy performance and energy conservation in health care buildings in Hellas", *Energy Conversion and Management*, 35, pp. 293-305.

[6] Kolokotsa, D., Tsoutsos, Th. & Papantoniou, S. (2012). "Energy conservation techniques for hospital buildings", *Advances in Building Energy Research*, 6(1), pp. 159-172. https://doi.org/10.1080/17512549.2012.672007

[7] Santamouris, M. & Argiriou, A. (1994). "Renewable energies and energy conservation technologies for buildings in Southern Europe", *International Journal of Solar Energy*, 15, pp. 69-79.

[8] Sofronis, H. & Markogiannakis, G. (2000). "Energy consumption in public hospitals", *Bulletin of Association of Mechanical Engineers*, Retrieved on 2/5/2018 from www.thelcon.gr/pdfs/publication%20hospitals.pdf. (In Greek).

[9] "Saving energy with energy efficiency in hospitals", CADDET 1997. Report retrieved on 2/5/2018 from https://www.certh.gr/dat/834E8024/file.pdf

[10] Van Schijndel, A.W.M. (2002). "Optimal operation of a hospital power plant", *Energy and Buildings*, 34(10), pp. 1055-1065. https://doi.org/10.1016/S0378-7788(02)00027-0

[11] Bujak, J. (2010). "Heat consumption for preparing domestic hot water in hospitals", *Energy and Buildings*, 42(7), pp. 1047-1055. DOI:10.1016/j.enbuild.2010.01.017

[12] Short, C.A. & Al-Maiyah, S. (2009). "Design strategy for low energy ventilation and cooling of hospitals", *Building Research and Information*, 37(3), pp. 264-292. doi: 10.1080/09613210902885156

[13] Saidur, R., Hasanuzzaman, M., Yogeswaran, S., Mohammed, H.A. & Hossain, M.S. (2010). "An end-use energy analysis in a Malaysian public hospital", *Energy*, 35(12), pp. 4780-4785. doi: 10.1016/j.energy.2010.09.012

[14] Herrera, A., Islas, J. & Arriola, A. (2003). "Pinch technology application in a hospital", *Applied Thermal Engineering*, 23(2), pp. 127-139. https://doi.org/10.1016/S1359-4311(02)00157-6

[15] Argiriou, A., Asimakopoulos, D., Balaras, C., Daskalaki, E., Lagoudi, A., Loizidou, M., Santamouris, M. & Tselepidaki, L. (1994). "On the energy consumption and indoor air quality in office and hospital buildings in Athens, Hellas", *Energy and Conversion Management*, 35, pp. 385-394.

[16] "Towards zero carbon hospitals with renewable energy systems", An Intelligent Energy Europe project, https://ec.europa.eu/energy/intelligent/projects/en/projects/res-hospitals

[17] Paksoy, H.O., Andersson, O., Abaci, S., Evliya, H. & Turgut, B. (2000). "Heating and cooling of a hospital using solar energy coupled with seasonal thermal energy storage in an aquifer", *Renewable Energy*, 19(1-2), pp. 117-122.

[18] Buonomano, A., Calise, F., Ferruzi, G. & Vanoli, L. (2014). "A novel renewable poly-generation system for hospital buildings: Design, simulation and thermo-economic optimization", *Applied Thermal Engineering*, 67, pp. 43-60. http://dx.doi.org/10.1016/j.applthermaleng.2014.03.008

[19] Vourdoubas, J. (2015). "Creation of zero CO_2 emission hospitals due to energy use. A case study in Crete, Greece", *Journal of Engineering and Architecture*, 3(2), pp. 79-86. doi: 10.15640/jea.v3n2a9

[20] Vourdoubas, J. (2016). "Possibilities of using solid biomass for covering the heating needs in hospitals in Crete, Greece", *Studies in Engineering and Technology*, 3(1), pp. 124-131. doi:10.11114/set.v3i1.1777

[21] Steins, Ch. & Zarrouk, S.J. (2012). "Assessment of a geothermal space heating system at Rotorua hospital, New Zealand", *Energy Conversion and Management*, 55, pp. 60-70.

[22] Tsoutsos, Th., Aloumpi, E., Gkouskos, Z. & Karagiorgas, M. (2010). "Design of a solar absorption cooling system in a Greek hospital", *Energy and Buildings*, 42(2), pp. 265-272. doi:10.1016/j.enbuild.2009.09.002

[23] Vanhoudt, D., Desmedt, J., Van Bael, J., Robeyn, N. & Hoes, H. (2011). "An aquifer thermal storage system in a Belgian hospital: Long term experimental evaluation of energy and cost savings", *Energy and Buildings*, 43(12), pp. 3657-3665. doi:10.1016/j.enbuild.2011.09.040

[24] Bizzari, G. & Morini, G.L. (2006). "New technologies for an effective energy retrofit of hospitals", *Applied Thermal Engineering*, 26, pp. 161-169. doi:10.1016/j.applthermaleng.2005.05.015

[25] Bizzari, G. & Morini, G.L. (2004). "Greenhouse gas reduction and primary energy savings via adoption of a fuel cell hybrid plant in a hospital", *Applied Thermal Engineering*, 24, pp. 383-400. doi:10.1016/j.applthermaleng.2003.09.009

[26] Renedo, C.J., Ortiz, A., Manana, M., Silio, D. & Perez, S. (2006). "Study of different co-generation alternatives for a Spanish hospital center", *Energy and Buildings*, 38(5), pp. 484-490. https://doi.org/10.1016/j.enbuild.2005.08.011

[27] Ziher, D. & Poredos, A. (2006). "Economics of a tri-generation system in a hospital", *Applied Thermal Engineering*, 26(7), pp. 680-687. doi:10.1016/j.applthermaleng.2005.09.007

[12] Energy consumption and carbon emissions in an Academic Institution in Greece. Can it become carbon neutral?

1. Introduction

Promotion of environmental sustainability is of paramount importance for coping with the current degradation of earth's environment. Academic Institutions should be in the forefront of global efforts to achieve the seventeen (17) UN's Sustainable Development Goals trying to offer a good example to their students as well as to local societies improving first of all their own sustainability. Current research investigates the possibility of zeroing net carbon emissions due to energy use in an Academic Institution located in Crete, Greece. This can be achieved by increasing its energy efficiency and by replacing fossil fuels use with renewable energies. Various mature, reliable and cost-effective renewable energy technologies can be utilized in the institute zeroing its net carbon emissions and footprint. Solar energy, solid biomass and low enthalpy geothermal energy are abundant in Crete and they are currently used in many sectors for heat, cooling and electricity generation. Achievement of this goal in the Institution is important since it reduces fossil fuels use as well as its carbon emissions complying with global efforts for climate change mitigation as well as with EU's goals to become carbon neutral continent by 2050.

2. Literature survey

Universities consume energy usually derived from fossil fuels and emit GHGs. Due to their pivotal role in society they should be in the forefront of global efforts to mitigate climate change. Although various studies concerning specific energy consumption in Academic Institutions have been implemented so far worldwide there is lack of reports concerning their transformation to carbon neutral organizations. Achievement of net zero carbon emission Institutions require replacement of fossil fuels and energy derived by fossil fuels with renewable energy sources. Use of various renewable energy technologies should be technically feasible and economically viable. It is necessary and challenging for Universities to investigate the possibility of reducing or zeroing their net carbon emissions due to operating energy use in the climate change era. This will facilitate investments in various renewable energy technologies for energy generation in their premises and reduction in their carbon emissions.

Khoshbankht et al, 2018 reported on energy use characteristics in higher education buildings. The authors have analyzed data from 80 University campus buildings in Australia. They stated that buildings used for research had the highest annual energy consumption at 216 KWh/m^2 while buildings used for offices had the lowest consumption at 137 KWh/m^2. Bernardo et al, 2018 estimated the energy saving potential in higher education buildings with reference in the building stock of Polytechnic Institute of Leiria, Portugal. The authors categorized Academic buildings according to their typology and they implemented energy audits in them. Their results indicated that potential energy savings vary between 10-34% in final energy consumption. Lo, 2013 reported on energy conservation in China's higher educational Institutes. The author investigated energy use in eight (8) Higher Education Institutes (HEIs) in China. He mentioned that HEIs have implemented non-technical interventions to save electricity. However, conservation of heat energy was not detected. The author concluded that due to limited financial resources realization of energy saving investments was rather limited. Ma et al, 2015 reported on energy consumption in University campus buildings. The authors stated that energy consumption in various Universities varies significantly. Annual electricity consumption varies from 2,000 KWh/person up to 40,000 KWh/person. Annual energy consumption per unit surface varies between 250-800 KWh/m^2 while annual electricity consumption varies between 80-200 KWh/m^2. Carbon emissions vary from 1.4-4 ton/person in European and Asian Universities while up to 6-8 ton/person in American Universities. Gul et al, 2015 reported on the relation of occupancy patterns in Academic buildings and their energy consumption. Analyzing the electricity demand data at regular intervals and the occupancy patterns in an Academic building they did not detect any significant correlation. They stated that the Academic building was controlled by a building energy management system which was independent from occupancy patterns. Alshuwaikhat et al, 2008 reported on integrated approaches for achieving campus sustainability. The authors stated that systematic and sustainable approaches for reducing environmental degradation in Academic buildings are generally lacking. The authors concluded that in order to be sustainable a University campus it must preserve the environment, stimulate economic growth and improve society. Boonthum, 2018 studied energy consumption in a public library building. He mentioned that the share of air conditioning in total energy use was at 81.49%, of lighting at 10.89 % and of other electric equipment at 7.62 %. Annual energy consumption in the library was 55.6 KWh/m^2 or 23.7 KWh/person using the library. Berrardi, 2015 reported on building energy consumption in US, EU and BRIC countries. The author stated that measures adopted in developed countries are insufficient to guarantee a significant reduction in energy consumption in their buildings. He also mentioned that energy consumption in BRIC countries has overcome total energy consumption in developed countries. Hinge et al, 2004 studied the energy use in commercial buildings around the world. The authors stated that there is a trend towards homogenization of the style and types of construction towards more similar "Western style" commercial buildings. They mentioned that annual energy consumption varies between 216 KWh/m^2 and 294 KWh/m^2 with an average value at 265 KWh/m^2. Energy consumption guidelines for higher education buildings in United Kingdom have been reported. Targeted annual energy consumption in higher education buildings varies significantly according to building type. It is lower in offices and teaching rooms while it is higher in catering buildings. Benchmarking values for annual electricity consumption are at 75 KWh/m^2 while for fossil fuels consumption at 182 KWh/m^2. Ouf et al, 2017 analyzed the energy consumption in school buildings in Manitoba, Canada. The authors stated that the age of school buildings had a statistically significant effect on their energy consumption with newer schools consumed less gas and more electricity than older and middle age schools. They also mentioned that middle-aged school buildings were the largest energy consumers. Vagi et al, 2011 analyzed the energy performance of secondary school buildings in northern Greece. The authors studied the energy consumption in ten high schools in Prefecture of Evros, Greece during 2001-2005. They stated that their mean annual heat energy consumption was at 81.50 KWh/m^2 while their mean annual electricity consumption was at 14.10 KWh/m^2. A toolkit for greening Universities, 2013 has been reported. The study stated that primary focus of a University's climate action plan should be on energy management. Energy management includes energy conservation, increase in energy efficiency and promotion of renewable and alternative energy sources including low carbon energy technologies. Vourdoubas, 2015 studied the creation of zero carbon emissions hospitals in Crete, Greece. The author stated that their annual energy consumption was at 366 KWh/m^2. He mentioned that combined use of various locally available renewable energy sources for covering all energy needs in hospitals could zero their net carbon emissions. Vourdoubas, 2017 reported on creation of net zero carbon emissions office buildings due to energy use in Crete, Greece. The author stated that their annual energy consumption was at 186 KWh/m^2. He indicated that use of solar thermal energy, solar photovoltaic energy, solid biomass and high efficiency heat pumps could cover all the energy needs in office buildings in Crete, Greece zeroing their net carbon footprint. Deshko et al, 2013 reported on estimation of energy performance in University campuses in Ukraine. The authors stated that annual total energy consumption varies between 31.6 KWh/m^2 and 608 KWh/m^2 with a mean value at 174 KWh/m^2. Escobedo et al, 2014 reported on energy consumption and GHG emissions scenarios in a University campus in Mexico. The authors stated that the covered area in the campus was 1.3 km^2 and it was using electricity, LPG and diesel oil. Total annual energy consumption was at 83 KWh/m^2 while lighting had the highest percentage in total energy use at 28%. They also mentioned that energy retrofitting of campus buildings could reduce energy consumption by 7.5% and CO_2 emissions by 11.3%. Luo et al, 2017 estimated the energy consumption in the library building in North China University of Science and Technology. The authors stated that the specific annual energy consumption was at 174 KWh/m^2 while lighting in the building had the highest share in energy consumption. Altan et al, 2014 implemented an energy performance analysis of Academic buildings in Sheffield University, UK. The authors stated that annual specific energy consumption of gas was at 158.74 KWh/m^2 while for electricity at 75.78 KWh/m^2. Comparing their results with existing data they mentioned that buildings performed poorly in terms of gas consumption while electricity consumption was satisfactory. Guan et al, 2016 reported on energy usage in a Norwegian University campus. The authors analyzed energy data during the last six years and they estimated the annual heat energy consumption at 99 KWh/m^2 and the annual electricity consumption at 206 KWh/m^2. Scheuer et al, 2003 reported on life cycle analysis of a new University building located in Michigan University campus. The authors mentioned that the covered surface of the building was 7,300 m^2 with a life span at 75 years. They have estimated the life cycle energy consumption at 316 GJ/m^2 or 4.21 GJ/m^{2}year (1,170 KWh/m^{2}year). The share of embodied energy in the life cycle energy consumption was estimated at 2.2% during construction and 0.2% during demolition. Electricity and HVAC consumption accounted at 94.4% and water heating at 3.3% of the life cycle energy use. Chung et al, 2014 reported on potential opportunities for energy conservation in existing buildings in a University campus in Korea. The authors estimated the annual energy intensity in buildings at 223 KWh/m^2 compared with an average value for Korean University buildings at 210 KWh/m^2. They also estimated the potential for energy conservation in University buildings in the range of 6%-30%. Zhou et al, 2013 reported on energy consumption and energy conservation in University buildings in Guangdong province, China. The authors mentioned that there is a great difference in energy intensity between different types of Universities. Analyzing data from ten Universities over a period of five years, they found that annual electricity consumption varies between 12 KWh/m² and 35.67 KWh/m² while annual gas consumption varies between 0.04 M³/M² and 1.02 M³/M². They also stated that there is great potential for energy saving and reduction of CO_2 emissions by 25%. Annual energy consumption in various University buildings around the world varies broadly according to existing literature from 83 KWh/m² up to 1,170 KWh/m² as it is presented in Table 1.

Table 1. Specific energy consumption in University buildings in various countries

Country	Source (Author, Year)	Specific annual energy consumption (KWh/m²)	Reference building
Australia	Khoshbanht et al, 2018	137-216	University campuses
Worldwide	Ma et al, 2015	250-800	University campuses
Thailand	Boonthum, 2018	55.6	University library building
UK	Energy efficiency best practice program	257	Benchmarking value for Universities
Ukraine	Deshko et al, 2013	174	University campuses
Mexico	Deshko et al, 2013	83	University campuses
North China	Luo et al, 2017	174	University library building
UK	Altan et al, 2014	158.74	University campuses
Norway	Guan et al, 2016	305	University campuses
USA	Scheuer et al, 2003	1,170	University campuses
South Korea	Chung et al, 2014	223	University

				campuses
Crete, Greece	Current work, 2018		164.96	University campuses

The aims of current work are:
1.   The estimation of specific energy consumption and carbon emissions due to operating energy use in an Academic Institute located in Chania, Crete, Greece,
2.   The description of renewable energy technologies which can be used in the Academic Institute for reducing or zeroing its net carbon emissions due to operating energy use, and
3.   The sizing and estimation of the investment cost of the required sustainable energy systems covering all energy loads in the Institution.

3. Description of the Academic Institute

The Academic Institute is a small size higher education and research public organization. Its premises are located in the suburbs of city of Chania including:
a)   Lecturing rooms and offices,
b)   Various research laboratories,
c)   A conference center,
d)   Dormitories for the students,
e)   A library,
f)   A catering facility, and
g)   Various traditional residential stone houses used by the participants in the conferences as well as by visiting professors.

Most of its buildings are more than thirty years old constructed with old building codes while they have inefficient thermal insulation. Although the local climate is mild, buildings require a lot of energy for their heating and cooling.

4. Energy consumption in the Institute

Energy sources which are currently used in the Institute include:
1.   Grid electricity for lighting, operation of various devices, electric machinery, apparatus and powering the heat pumps used in air-conditioning,
2.   Diesel oil for space heating and domestic hot water production,
3.   LPG for cooking,
4.   Solar thermal energy with thermo-siphonic systems with flat plate collectors producing DHW,

5. Solar photovoltaics (solar-PVs) generating electricity with nominal power 13.5 KW$_p$, and
6. Solid biomass in a wood stove for space heating.

Climate conditions in the premises of the Institute are presented in Table 2.

Table 2. Climate conditions in Chania, Crete, Greece

Parameter	Value
Latitude	35°. 52'. 1"
Mean annual air temperature	18.1 °C
Annual precipitation	550 mm
Mean annual humidity	67 %

Various fuels and energy sources used in the Institute are presented in Table 3.

Table 3. Fuels and energy sources used annually in the Institute[1]

Energy source	Quantity	Energy content (KWh/year)	% of total energy	Energy/fuel Cost (€/year)
Electricity	1,569,597 KWh$_{el}$	1,569,597	84.96	104,770
Diesel oil[2]	9,259 Lt	97,775	5.29	8,539
LPG[3]	22,224 Lt	154,235	8.35	14,522
Solid biomass[4]	300 Kg	1,463	0.08	114
Solar-PV	20,200 KWh$_{el}$	20,200	1.09	0
Solar thermal	4,300 KWh$_{th}$	4,300	0.23	0
Total		1,847,570	100	127,945

[1] Average values for years 2016 and 2017, [2] Heating value of diesel oil, 10.56 KWh/lt [3], Heating value of LPG, 6.94 KWh/lt, [4]Heating value of biomass, 4,200 Kcal/kg

5. Carbon emissions due to operating energy use in the Institute

Use of fossil fuels and electricity generated by fossil fuels in the Institute results in CO_2 emissions. Carbon dioxide emissions due to fossil fuels and electricity generated by fossil fuels are presented in Table 4.

Table 4. Annual carbon dioxide emissions due to conventional energy use in the Institute

Energy used	Annual CO$_2$ emissions (kgCO$_2$)	Specific annual CO$_2$ emissions[4] (kgCO$_2$/m^2)	Annual CO$_2$ emissions[5] (kgCO$_2$/employee)
Electricity[1]	1,177,198	105.11	16,817
Diesel oil[2]	26,399	2.36	377
LPG[3]	35,474	3.17	507
Total	1,239,071	110.64	17,701

[1]*Electricity generation, 0.75 kgCO$_2$/KWh,* [2] *Diesel oil, 0.27 kgCO$_2$/KWh,* [3] *LPG, 0.23kgCO$_2$/KWh,* [4]*Covered area in the Institute: 11,200 m^2,* [5]*Number of employees: 70*

6. Use of renewable energy technologies for energy generation in the Institute

Various locally available renewable energy sources can be used on-site or off-site for electricity, heat and cooling generation, including:

a)   Solar thermal energy with solar thermo-siphonic systems with flat plate collectors for DHW production,

b)   Solar-PV energy for electricity generation,

c)   High efficiency ground source heat pumps (GSHPs) for heating and cooling.

The abovementioned renewable energy technologies are broadly used in Crete providing heat, cooling and electricity. They are mature, reliable and well proven technologies. Their cost-effectiveness has been established while they are accepted by local societies. Combined properly they could cover all the energy requirements in the Institute zeroing the use of fossil fuels as well as carbon emissions in its premises. The average annual wind speed in the Institute is low and installation of wind turbines is not indicated. Solid biomass could be also used for heat generation. Olive kernel wood is a by-product of olive oil processing industry which is currently used in Crete for heat generation in buildings, in industry and in agriculture. Its price compared with its heating value is attractive. However, during its burning flue gases are produced and use of alternative heat generation technologies is preferable. Solar thermal cooling could be also used for space cooling. However, this technology has not been commercialized so far in Crete and use of GSHPs is rather preferable.

7. Requirements for a carbon neutral Institution

In order to zero net carbon emissions due to operating energy use in the Institute the following criteria should be fulfilled:

1. All fossil fuels used for heat generation should be replaced with locally available renewable energy sources, and
2. All the grid electricity used in the Institute should be annually offset with green electricity, preferably solar-PV electricity.

Use of solar-PVs connected into the grid is allowed in buildings and enterprises according to net-metering regulations which are in force for the last four years in Greece (Greek Ministerial Decree, 2014). Although there are limitations regarding the maximum power allowed in solar-PVs installed in buildings with net metering regulations in Greek islands it is expected that restrictions will be removed after the interconnection of Crete's and continental Greece's grids. It should be noted that the share of fossil fuels in grid electricity generation in Crete is approximately 80% while the remaining 20% is derived from renewable energies mainly by solar-PVs and wind parks. Therefore, offsetting annually grid electricity consumption in the Academic Institute with green solar electricity will result in negative carbon emissions in it.

8. Sizing various sustainable energy systems covering all energy requirements

8.1 Sustainable energy technologies use

The following sustainable energy technologies are considered for covering all energy demand in the Academic Institute

1. Solar thermal technology with thermo-siphonic systems and flat plate collectors. It is assumed that they can provide 50% of the annual requirements in DHW production,
2. High efficiency GSHPs with COP equal at 3.5 covering the requirements in space heating and cooling. They can additionally cover the remaining 50% requirements in DHW production,
3. Solar-PV modulus providing annually all the electricity required for lighting, operation of electric equipment and powering the GSHPs. Solar electricity generation can offset annually all the grid electricity consumed in the Institute.

8.2 Estimation of energy loads

In order to estimate the energy loads required in the Institute the following assumptions have been made:
1. All the equipment in the kitchen using LPG will be replaced with electric equipment,

2. Half of the diesel oil used in the Institute will be consumed in space heating while the rest in DHW production, and
3. 40% of the total electricity use is consumed in lighting, operation of various equipment and devices while the rest 60% for powering the GSHPs.

Energy consumption for heating, cooling, DHW production, lighting and operation of various electric devices in the Institute is presented in table 5.

Table 5. Energy consumption for heating, cooling, DHW production, lighting and operation of various electric devices in the Institute

Energy use	Energy content (KWh)	% of total energy use
Electricity use for lighting and operation of electric devices	703,200	38.83
Electricity use in GSHPs	1,054,800	58.23
DHW production	53,187.5	2.94
Total	1,811,187.5	100

8.3 Sizing the necessary sustainable energy systems

For sizing the sustainable energy systems required for covering all energy needs in the Academic Institute the following assumptions have been made:
1. Energy consumption in the Institute will remain the same without any improvements in energy efficiency,
2. The annual electricity generation by a solar-PV system in Crete is 1,500 KWh_{el}/KW_p,
3. The annual heat generation by a solar thermo-siphonic system with flat plate collectors in Crete is 500 KWh_{th} per m^2 of collectors, and
4. In order to cover the peak heat and cooling loads the size of the GSHPs will be double than the size required for covering the average loads.

The size of sustainable energy systems is presented in table 6.

Table 6. Size of sustainable energy systems used for covering all the electricity needs in an Academic institution located in Crete, Greece

Sustainable energy System	Size
Solar thermal	106.4 m^2 of flat plate solar collectors
Solar-PV	1,172 KW_p
GSHPs	424 KW_{el}

8.4 Capital cost of sustainable energy systems

The following assumptions have been made for the estimation:
1. The capital cost of the solar-PV system is 1,300 €/KW$_p$,
2. The capital cost of the solar thermal system is 300 €/m^2 of collectors, and
3. The capital cost of the GSHPs is 1,200 €/KW$_{el}$.

The capital cost of sustainable energy systems in the Institute is presented in table 7.

Table 7. Capital cost of the sustainable energy systems

System	Cost (€)	Cost (€ per m^2 of covered surface in the Institute)
Solar-PV	1,523,600	136.04
Solar thermal	31,920	2.85
GSHPs	508,800	45.43
Total	2,064,320	184.32

9. Discussion and conclusions

The main energy source currently used in the Academic Institute in Crete, Greece is electricity accounting for approximately 85% of its total energy consumption while renewable energies contribute only 1.4% in its annual energy balance. Among renewable energies, solar thermal energy, solar-PV energy and solid biomass are currently used for heat and electricity generation but their contribution to the total energy balance in the Institute is rather negligible. Specific total annual energy consumption has been estimated at 164.96 KWh/m^2 which is lower than the benchmarking value for British universities which is 257 KWh/m^2 while its annual carbon emissions are 110.64 kgCO$_2$/m^2. Solar energy is abundant in Crete, Greece and it is already used for heat and power generation. Combined use of solar thermal energy, solar-PV energy and GSHPs could cover all the energy requirements in the Academic Institute located in Crete, Greece zeroing its net carbon emissions due to energy use. The abovementioned technologies are mature, reliable, cost-effective and well proven. The legal framework for offsetting grid electricity consumption with solar-PV electricity exists already in Greece allowing on-site or off-site installation of photovoltaic modulus. A preliminary calculation of the size of sustainable energy systems required has been made and their capital cost has been estimated at 184.32 € per m^2 of the Institute's covered surface. In the previous analysis improvement of the energy efficiency in the Institute has not been foreseen. If though the energy efficiency will be improved reducing its annual specific energy consumption then the size of sustainable energy systems required would be smaller. Their overall capital cost at 2,064,320 € is equal to around fifteen times the current annual cost of energy consumption in the Institute. The size of the solar thermal system required is small with an area of flat plate collectors at 106.4 m^2 and it can be installed on the terrace of existing buildings. The size though of the solar-PV system is large and its installation requires a large surface which is not available in the Institute. However, it can be installed outside of the Institute's premises. If financial support regarding the capital cost of the energy systems will be offered by the government the investments would be profitable. Current study indicates that sustainable energy technologies required for creating a carbon neutral Academic Institute in Crete, Greece are mature, reliable and commercially used. They could be also economically attractive if a financial subsidy was offered. Further work should be focused in examining a) the possibility of using solid biomass burning for heat generation and solar thermal cooling for space cooling, and b) low carbon energy technologies like co-generation systems and fuel cells in the Institute.

References

[1] Alshuwaikhat, H.M. & Abubakar, I. (2008). "An integrated approach to achieving campus sustainability: assessment of the current campus environmental management practices", *Journal of Cleaner Production,* 16, pp. 1777-1785. doi:10.1016/j.jclepro.2007.12.002

[2] Altan, H., Douglas, J.S. & Kim, Y.K. (2014). "Energy performance analysis of University buildings: Case studies at Sheffield University, U.K.", *Journal Architectural Engineering Technology,* 3(3), pp. 129. doi:10.4172/2168-9717.1000129

[3] Bernardo, H. & Oliveira, F. (2018). "Estimation of energy saving potential in higher educational buildings supported by energy performance benchmarking: A case study", *Environments,* 5, 85, pp. 1-14. doi:10.3390/environments5080085

[4] Berrardi, U. (2015). "Building energy consumption in US, EU and BRIC countries", *Procedia Engineering,* 118, pp. 128-136. doi: 10.1016/j.proeng.2015.08.411

[5] Boonthum, E. (2018). "Study on energy consumption in library building", *Kasem Bundit Engineering Journal,* 8(2), pp. 117-130.

[6] Chung, M.H. & Rhee, E.K. (2014). "Potential opportunities for energy conservation in existing buildings on University campus: A field survey in Korea",

Energy and Buildings, 78, pp. 176-182. http://dx.doi.org/10.1016/j.enbuild.2014.04.018

[7] Deshko, V.I. & Shevchenko, O.M. (2013). "University campuses energy performance estimation in Ukraine based on measurable approach", *Energy and Buildings*, 66, pp. 582-590. http://dx.doi.org/10.1016/j.enbuild.2013.07.070

[8] "Energy efficiency in further and higher education, Cost effective low energy buildings, Energy consumption guide 54". Retrieved at 12/11/2018 from http://www.cibse.org/getmedia/f944f3e0-e047-4c62-af85-43f02533f2de/ECG54-Energy-Use-in-Further-Higher-Education-Buildings.pdf

[9] Escobedo, A., Briceno, S., Juarez, H., Castillo, D. Imaz, M. & Sheinbaum, C. (2014). "Energy consumption and GHG emission scenarios of a University campus in Mexico", *Energy for Sustainable Development*, 18, pp. 49-57. http://dx.doi.org/10.1016/j.esd.2013.10.005

[10] "Greening Universities toolkit, Transforming Universities into green and sustainable campuses", United Nations Environmental Program 2013. Retrieved at 12/11/2018 from https://europa.eu/capacity4dev/unep/document/greening-universities-toolkit-transforming-universities-green-and-sustainable-campuses

[11] Guan, J., Nord, N. & Chen, S. (2016). "Energy planning of University campus building complex: Energy usage and coincidental analysis of individual buildings with a case study", *Energy and Buildings*, 124, pp. 99-111. http://dx.doi.org/10.1016/j.enbuild.2016.04.051

[12] Gul, M.S. & Patidar, S. (2015). "Understanding the energy consumption and occupancy of a multi-purpose Academic building", *Energy and Buildings*, 87, pp. 155-165. http://dx.doi.org/10.1016/j.enbuild.2014.11.027

[13] Hinge, A., Bertoldi, P. & Waide, P. (2004). "Comparing commercial building energy use around the world". Retrieved at 7/11/2018 from https://aceee.org/files/proceedings/2004/data/papers/SS04_Panel4_Paper14.pdf

[14] Khoshbakht, M., Gou, Z. & Dupre, K. (2018). "Energy use characteristics and benchmarking for higher education buildings", *Energy and Buildings*, 164, pp. 61-76. https://doi.org/10.1016/j.enbuild.2018.01.001

[15] Lo, K. (2013). "Energy conservation in China's higher education Institutions", *Energy Policy*, 56, pp. 703-710. http://dx.doi.org/10.1016/j.enpol.2013.01.036

[16] Luo, R. Han, Y. & Zhou, X. (2017). "Characteristics of campus energy consumption in North China University of Science and Technology", *Procedia Engineering*, 205, pp. 3816-3823. 10.1016/j.proeng.2017.10.098

[17] Ma, Y.T., Lu, M.Y. & Weng, J.T. (2015). "Energy consumption status and characteristics analysis of University campus buildings", *5th International Conference on Civil Engineering and Transportation*, (ICCET 2015), (p.p. 1240-1243).

[18] Ministerial Decree, Greek Ministry of Environment, Energy and Climatic Change (ΥΡΕΚΑ), 30/12/2014, ΑΠΕΗΛ/Α/Φ1/οικ.24461, (ΦΕΚΒ' 3583/31.12.2014) (In Greek).

[19] Ouf, M.M. & Issa, M.H. (2017). "Energy consumption analysis of school buildings in Manitoba, Canada", *International Journal of Sustainable Built Environment*, 6, pp. 359-371. http://dx.doi.org/10.1016/j.ijsbe.2017.05.003

[20] Scheuer, Ch., Keoleian, G.A. & Reppe, P. (2003). "Life cycle energy and environmental performance of a new University building: modeling challenges and design implications", *Energy and Buildings*, 35, pp. 1049-1064. doi:10.1016/S0378-7788(03)00066-5

[21] Vagi, F. & Dimoudi, A. (2011). "Analyzing the energy performance of secondary schools in Northern Greece", in *World Renewable Energy Congress 2011*, 8-13 May 2011, Linkoping, Sweden.

[22] Vourdoubas, J. (2015). "Creation of zero CO_2 emission hospitals due to energy use. A case study in Crete, Greece", *Journal of Engineering and Architecture*, 3(2), pp. 79-86. doi: 10.15640/jea.v3n2a9

[23] Vourdoubas, J. (2017). "Creation of zero CO_2 emission office buildings due to energy use: A case study in Crete, Greece", *International Journal of Multidisciplinary Research and Development*, 4(2), pp. 165-170.

[24] Zhou, X., Yan, J., Zhu, J. & Cai, P. (2013). "Survey of energy consumption and energy conservation measures for colleges and Universities in Guangdong province", *Energy and Buildings*, 66, pp. 112-118. http://dx.doi.org/10.1016/j.enbuild.2013.07.055

[13] Possibilities of creating net zero carbon emissions cultural buildings. A case study of the museum in Eleftherna, Crete, Greece

1. Introduction

Global efforts for mitigating climate change have increased efforts to reduce energy consumption and carbon emissions in public buildings including cultural heritage buildings. Improving energy efficiency and reducing carbon emissions in museums requires a compromise among various factors including: a) Preservation of architectural characteristics of historic buildings when integrating benign energy technologies, b) Maintaining the required indoor climate conditions for protection of the artifacts exhibited, and c) Achieving comfort for visitors and absence of indoor pollutants. In particular integration of solar photovoltaic (solar-PV) systems or other solar energy technologies in museum buildings should maintain their historic characteristics. Various approaches have been used for improving energy efficiency in museum buildings, including: a) Altering the set-points of indoor temperature and relative humidity without damaging the artifacts exposed, b) Using passive and active energy saving techniques and technologies combined with building retrofitting, and c) Installing renewable energy technologies in them. Zeroing the net carbon footprint due to energy use in museum buildings assists in mitigation of climate change and in achievement of the EU target for carbon neutral Europe by 2050.

2. Literature survey

2.1 Energy consumption in museums and historical buildings

A case study on energy consumption and CO_2 emissions in two museums in Hangzhou, China has been reported [1]. The authors estimated the specific annual energy consumption during the operation of two museums at 176.4 KWh/m² and at 91 KWh/m² while their carbon emissions were estimated at 167.58 kgCO$_2$/m² and at 86.45 kgCO$_2$/m² correspondingly. They have also mentioned that during their operation phase the two museums consumed 94-96.5% of their life cycle energy consumption while they emitted 90-94% of their life cycle CO_2 emissions. Strategies regarding energy saving in museum's air-conditioning have been reported [2]. The authors implemented a case study in a modern museum and stated that energy savings up to 40% can be achieved if the indoor relative humidity (RH) changes from 50+/-2% to 50+/-10%. They also mentioned that smaller energy savings at 6-13% can be achieved if the indoor temperature varies between 21°C in winter and 23°C in summer instead of being constant at 22°C. An energy study in the museum of Science and Industry located in Chicago, USA has been presented [3]. The authors stated that the museum was using electricity and natural gas for covering its energy requirements. The share of electricity in lighting was 30%, in space cooling 22%, and the rest 48% in the operation of electric equipment. The share of natural gas in space heating was at 96% and in hot water production at 4%. A study on energy efficiency in retrofitted and new museum buildings in Europe has been presented [4]. The authors have studied eight museums in Europe and they estimated the total annual energy consumption in five of them. It was varying from 69 KWh/m^2 to 149 KWh/m^2. They mentioned that using energy saving techniques the overall energy consumption in museums can be reduced from 39 % to 77%. The authors additionally stated that the payback period of energy saving investments was varying between 11.3 and 49 years. A study on the increase of energy efficiency in cultural buildings has been reported [5]. The authors investigated the improvement of energy efficiency in nine European museums. They tried to achieve total energy savings by 35-40% and to reduce CO_2 emissions by 50%. Studying the Greek museums in Delphi and Heraklion they mentioned that use of passive solar techniques and various energy saving measures can improve their energy efficiency without affecting their functionality. Energy consumption in Hermitage museum in Amsterdam has been reported [6]. The authors investigated the impacts of energy consumption on fluctuations of indoor climate. They stated that slight alteration in indoor temperature and RH can reduce the overall energy consumption in the museum by 49-63%. They also mentioned that keeping the indoor climate conditions constant at 21°C and 50% RH the annual energy consumption was at 1,053 KWh/m^2. When indoor climate conditions were slightly altered energy consumption was reduced at 385-534 KWh/m^2. A methodology for mapping future energy needs in European museums has been presented [7]. The authors developed a prediction method for future energy demand in different types of museum buildings and climate conditions all over Europe. They used seven performance indicators including indoor temperature, indoor RH, heating demand, cooling demand, humidification demand, dehumidification demand and total energy demand. A review on indoor environmental conditions in museums and their impact on energy consumption has been published [8]. The authors stated that maintaining the required environmental conditions for museum artifacts might not lead to the desired reduction in energy consumption. They have also mentioned that indoor environmental control includes the following parameters: temperature, RH, lighting and indoor pollutants. Energy conservation in museums using different set-point strategies has been reported [9]. The authors stated that many museums employ tight indoor climate restrictions in order to protect their artifacts from degradation. They have implemented a case study in a museum in Netherlands simulating its energy consumption at different set-point strategies. They concluded that optimum strategy of indoor climate yields energy savings at 77% improving thermal control and decreasing chemical degradation. A case study in Amsterdam museum regarding a more energy efficient Heat Ventilation Air Conditioning (HVAC) system has been presented [10]. The authors stated that maintaining a strict indoor climate reduces the risks for deterioration of the objects but increases the energy demand in the museum which was estimated at 84.55 KWh/m^2. They mentioned that performing computer simulations using less strict indoor climate control the initial annual energy demand of the HVAC system in the museum was reduced by 15%. A simulation study in four museums investigating the energy impact of different climate classes for the ASHRAE museum has been implemented [11]. The authors stated that significant energy savings are achieved when moving from class AA to class B which adequately protects most artifacts in museums. The amount of energy saved, they mentioned, depends on the type of the building and its characteristics. An investigation of possible energy efficient interventions in Galleria Borghese, Rome has been made [12]. The authors stated that improving energy efficiency in the air conditioning system, particularly in cooling during the summer, had the highest impact regarding its sustainability. A report on feasibility of energy retrofitting in historical buildings in Italy has been presented [13]. The authors mentioned that energy retrofitting in this type of buildings has to solve two important points: the restrictions regarding the use of renewable energy technologies in them and the high payback time of energy investments.

2.2 Use of renewable energies in museums

A report on energy efficient museum buildings has been made [14]. The author stated that museum buildings can be energy efficient although their performance requirements are ambitious. He indicated that their energy consumption can be reduced to less than one tenth compared to traditional museum buildings. The author mentioned that the Emil-Schumacher museum in Hagen had annual energy consumption at 117 KWh/m^2 which could be covered with the use of solar-PVs, geothermal heat pumps (GHPs) and earth to air heat exchangers. An investigation on the use of energy efficient approaches and integration of renewable energies in historical buildings including use of solar energy and GSHPs has been presented [15]. The authors stated that although the integration of solar energy in historical buildings is difficult due to lack of space and the need to preserve the exterior architecture there are many successful examples worldwide. Additionally, they mentioned, ground source heat pumps GSHPs are popular green energy systems improving energy efficiency in historic buildings. A report on the integration of solar energy technologies in historic buildings with reference to a case study in Katania, Italy has been published [16]. The authors stated that application of solar energy technologies in the external building envelope should comply with preservation requirements in historical buildings. They indicated that application of solar energy technologies in historic buildings should take into account their architectural, construction, energy and cultural aspects. An investigation on the performance of GSHPs in two historic buildings located in Venice and Florence, Italy has been made [17]. The authors compared, using computer simulation, the performance of a GSHP with a) an air source heat pump, and b) a conventional system with a gas boiler and air-to-water chiller. Their results indicated that the GSHP system is the best solution in terms of primary energy savings. A feasibility study concerning the use of a GSHP for heating and cooling in a historic building has been implemented [18]. The authors stated that the GSHP had the lowest annual energy requirements from non-renewable energy sources compared with other energy systems. A simplified method for assessing environmental and energy quality in museum buildings has been published [19]. The proposed evaluation method is based on three parameters including: a) Environmental performance evaluation, b) Energy performance evaluation, and c) Assessment of environmental and energy quality. The author stated that aspects impacting environmental and energy quality include heritage conservation, human comfort and energy efficiency. A case study on energy refurbishment of a historic building located in Perugia, Italy has been realized [20]. The building had total covered area at 7,000 m^2 and it was using natural gas for heating and a condensing external unit for cooling. The authors studied the replacement of the old heating and cooling system with a GSHP concluding that its annual heating and cooling consumption has been significantly reduced from 69 KWh/m^2 to 30 KWh/m^2. A report on improving energy efficiency and solar energy integration in historic buildings has been published [21]. The authors stated that in various building retrofitting projects, improving energy efficiency and using renewable energy technologies is not always possible without compromises. They also mentioned that use of solar energy technologies in historic buildings can be hindered due to preservation problems and aesthetic requirements. A study on improving energy efficiency in museums with reference to Musei Senesi in Torino, Italy has been implemented [22]. The authors stated that there are few cases where energy retrofitting in museums has been carried out and few data are available regarding their energy performance after retrofitting. They have realized a study in 43 museums producing a self-assessment check-list and a handbook. The authors suggested the use of the following renewable energy technologies in museum buildings: Solar-PV panels, solar thermal panels, biomass and heat pumps.

2.3 Zero carbon emissions buildings

A study on a zero carbon emissions science museum in Minnesota, USA has been implemented [23]. The museum has been designed to operate as zero carbon emissions building which covers annually all its energy requirements with renewable energies. The museum uses solar-PV panels for electricity generation and a GHP for air conditioning. The heat pump used electricity generated by the solar-PV system. A report on energy consumption and carbon emissions in an Academic Institution located in Crete, Greece has been published [24]. Its specific annual energy consumption has been estimated at 164.96 KWh/m² while its annual carbon emissions at 110.64 kgCO$_2$/m². The author mentioned that the combined use of solar thermal energy, solar-PV energy and GSHPs could cover all its annual energy requirements zeroing its net carbon emissions. He also estimated that the investment cost of all necessary sustainable energy systems was at 184.32 €/m² of its covered surface. A study on energy consumption and carbon emissions in Venizelio hospital in Crete, Greece has been realized [25]. The author estimated its annual energy consumption at 280.4 KWh/m² and its annual carbon emissions due to operating energy use at 168 KWh/m². He also stated that the combined use of solar thermal energy, solar-PV energy, solid biomass and GSHPs can cover all its annual energy requirements zeroing its net carbon emissions. Energy consumption in various museums according to literature reviewed is presented in Table 1.

Table 1. Energy consumption in museum buildings

Author/year	Location of the museum	Total annual energy consumption (KWh/m^2)	Annual energy consumption for HVAC (KWh/m^2)
Ge et al, 2015	Hangzhou, China (two museums)	176.4 and 91	
Zannis et al, 2006	Europe (average in five museums)	69-149	
Kompatscher et al, 2017	Europe, Netherland		84.55
Mueller, 2013	Europe, Germany	117	
Pisello et al, 2014	Europe, Italy		69

Source: Various authors

The aims of the current work are:
a) To indicate the reliable and cost effective renewable energy technologies which can cover all annual energy requirements in the museum of Eleftherna, Crete, Greece,
b) To estimate the size of the required renewable energy systems which can cover all the annual energy requirements in the museum of Eleftherna, Crete, Greece, and
c) To estimate CO_2 emissions savings achieved as well as the cost of the sustainable energy systems required.

Due to absence of accurate data on energy consumption in Eleftherna's museum various published data from other museum buildings in Europe and worldwide have been used in order to estimate its specific energy consumption and the size of the proposed renewable energy systems. For more accurate calculations reliable data on energy consumption during the last years in Eleftherna's museum must be used.

3. Requirements for zeroing net carbon emissions due to energy use

Museums consume energy during their operation for lighting, air-conditioning and operation of various electric equipments. Usually they consume grid electricity and/or natural gas. In order to zero their net carbon emissions due to operating energy use the following two criteria should be fulfilled:
1. They should replace all fossil fuels used with renewable energy sources, and

2. Grid electricity used annually should be compensated with solar-PV electricity generated preferably in situ and injected into the grid. This is allowed in many countries according to net-metering regulations.

Since fossil fuels are not used in the museum and grid electricity used, generated mainly from fossil fuels, is annually offset with green solar electricity its net carbon emissions are zeroed. It is assumed though that all grid electricity is generated by fossil fuels which is not true since part of it, in Crete, is generated by renewable energies mainly by solar and wind energy. It should be noted that the share of operating energy use to life cycle energy use in museums is higher than 90% while the share of embodied energy use is less than 10% [1].

4. The Archaeological Museum of Eleftherna in Crete, Greece

The museum of ancient Eleftherna [26] was inaugurated in 2016 and it is located in the Prefecture of Rethymno, Crete, Greece. It is an on-site museum displaying artifacts from the nearby archaeological site dating from 3,000 BC until 1,300 AD. It is the only museum in Crete exhibiting objects from the Homer era including the famous Tomb of Warriors dated at 830-730 BC. The museum is housed in a modern building with a covered area of 1,800 m^2. Using data reported in literature surveyed it is assumed that its specific annual energy consumption for HVAC is 75 KWh/m^2 while its specific total annual energy consumption is 120 KWh/m^2. Therefore, the total annual energy consumption in Eleftherna museum is estimated at 216,000 KWh/m^2 while its annual energy consumption for HVAC at 135,000 KWh/m^2.

5. Use of renewable energy technologies in the museum

Renewable energy technologies can be used in Eleftherna museum for covering all its annual energy requirements zeroing its net carbon emissions. Solar-PV electricity can be used for electricity generation and a high efficiency GSHP for covering all its heating and cooling needs. Solar-PV systems can generate all the electricity required annually for lighting, operation of electric machinery and operation of the heat pump. Solar electricity generated can annually offset all grid electricity consumption in the museum. Alternatively, solid biomass, locally produced in Crete, can be used for space heating although this fuel is not suggested for use in the museum. Solar thermal panels can be used for hot water production although its requirements in hot water are low. Space cooling can be also provided with a solar thermal cooling system. Currently though, solar thermal cooling technology is not mature, reliable and commercialized in Crete.

Since renewable energy technologies can provide all annual energy requirements in the museum of Eleftherna its net carbon emissions due to energy use could be zero. Renewable energy technologies which can be used in Eleftherna's museum are presented in Table 2.

Table 2. Renewable energy technologies which can be used in Eleftherna museum

Renewable energy technology	Energy generated
Solar-PV panels	Electricity
Solar thermal panels	Hot water
Solar thermal cooling	Space cooling
Solid biomass burning	Space heating
Geothermal heat pump	Space heating and cooling

Source: own estimations

5.1 Use of solar-PV energy for electricity generation

Solar-PV systems including crystalline or amorphous silicon modulus are broadly used today in Crete, Greece. They are used either in buildings offsetting grid electricity consumption with net-metering regulations or in individual systems generating electricity and injected into the grid with feed-in tariffs. For the climate conditions of Crete the average annual electricity generation with solar-PVs is estimated at 1,500 KWh/KW$_{peak}$. Generation of the electricity consumed annually in Eleftherna's museum at 216,000 KWh/m^2 requires a solar-PV system with a nominal capacity at 144 KW$_{peak}$. The necessary surface for installation of this solar-PV system is approximately 1,440 m^2. The solar system can be installed on the building's terrace or alternatively outside of the museum building probably on the roof of a car park.

5.2 Use of high efficiency heat pumps for air-conditioning

A high efficiency heat pump like a GSHP can be used for air-conditioning of Eleutherna's Museum building. Assuming that it will operate for 6 hours daily its electric power will be 61.6 KW. In order to cover the peak heating and cooling loads its electric power will be doubled at 123.2 KW.

6. Environmental and economic considerations

Installation of renewable energy systems in Eleftherna's museum covering all its annual energy needs results in zeroing its carbon emissions. Assuming that generation of 1 KWh of grid electricity with fossil fuels in Crete is equivalent with emissions of 0.75 KgCO$_2$ the total annual decrease in carbon emissions due to green electricity use in the museum is 162 tons. The cost of the required solar-PV system can be estimated taking into account that installation of 1 KW$_{peak}$ in Crete, Greece costs 1,750 €. The total cost of installing solar-PVs at 144 KW$_{peak}$ is 252,000 €. The cost of the GSHP can be calculated assuming that its unit cost is 2,200 €/KW. The cost of a heat pump at 123.2 KW will be 271,040 €. The size, cost and carbon emissions savings due to use of renewable energy systems in Eleftherna's museum are presented in Table 3.

Table 3. Size, cost and carbon emissions savings due to use of renewable energy systems in Eleftherna's museum

Renewable energy system	Size	Cost (€)	Cost per m² of covered surface (€)	Annual Carbon emissions savings (ton CO$_2$)	Annual carbon emissions savings per m² of covered surface (kgCO$_2$)
Solar-PV	144 KW$_{peak}$	252,000	140	162	90
GSHP	123.2 KW$_{el}$	271,040	150.58		
Total		523,040	290.58	162	90

Source: own estimations

7. Discussion

Studies related with museums which cover all their energy needs with renewable energy sources are rather limited. Most of existing studies are focused in the reduction of their energy consumption with various energy saving interventions and use of various renewable energy technologies for power generation and air-conditioning. Annual energy consumption in various museum buildings is lower than in other types of buildings according to existing studies. Financing the use of renewable energy technologies in museum buildings could be a problem particularly when public funding is limited due to various reasons. Alternatively, involvement of the private sector is a preferable option offering financial support through Energy Saving Companies (ESCOs) since energy investments are cost-effective and profitable. Estimation of the required renewable energy systems covering all the energy needs in of Eleftherna's museum have been made without considering any reduction in its energy consumption which probably can be achieved altering the set points in temperature and RH control of indoor environment. In this case the size and cost of sustainable energy systems required will be smaller. The proposed green energy technologies are mature, reliable and cost-effective used in many applications so far. Promotion of the above-mentioned sustainable energy systems in the museum presupposes: a) Their technical maturity, reliability and cost-effectiveness, b) The possibility of financing these energy investments, c) The existence of the appropriate legal framework allowing the use of these green energy systems and the injection of solar electricity into the grid, and d) The existence of public policies promoting the creation of nearly zero or zero carbon emissions public buildings complying with European directives and national legislation.

8. Conclusions

Museums utilize energy in order to maintain favorable indoor conditions preserving the exposed artifacts, offering comfort to visitors and minimizing indoor pollution. Use of sustainable energy technologies for reducing their carbon emissions should comply with the needs of preserving architectural characteristics in museum buildings. Museum of Eleftherna in Crete, Greece is a recently established modern museum which could zero its carbon emissions using renewable energy technologies generating all its annual energy requirements. Use of GSHPs covering all the annual air-conditioning needs has been studied combined with a solar-PV system generating all the annual electricity required. Annual energy consumption for air-conditioning in Eleftherna's museum with covered surface of 1,800 m^2, has been estimated at 135,000 KWh while its total annual energy consumption at 216,000 KWh. For covering all its air-conditioning requirements a geothermal heat pump at 123.2 KW is needed while a solar-PV system with nominal capacity at 144 KW$_{peak}$ could generate all annual electricity needs. The cost of the required GSHP has been estimated at 271,000 € while the cost of the solar-PV system at 252,000 €. The total cost of the green energy systems is 523,040 € or 290.58 € per m^2 of its covered surface. Total annual CO_2 emissions savings due to use of renewable energy systems in the museum have been estimated at 162 tonsCO_2 or 90 kgCO_2 per m^2 of its covered surface. Future work should be focused in monitoring the real annual energy consumption for lighting, air-conditioning and in other sectors in Eleftherna's museum. Additionally, possibilities of reducing its overall energy consumption without damaging the exposed artifacts as well as visitor's comfort should be studied. Further research regarding enegy consumption in other cultural buildings in Greece should be implemented in order to study and assist the promotion of renewable energy technologies in them as well as the reduction of their carbon footprint due to energy use. Deployment of sustainable energy investments in museum buildings requires awareness raising and sensitization of local and public authorities engaged with museum's operation, including their staff, regarding the benefits due to use of benign green energy sources in them.

References

[1] Ge, J., Luo, X., Hu, J. & Chen, S. (2015). "Life cycle energy analysis of museum buildings: A case study of museums in Hangzhou", *Energy and Buildings*, 109, pp. 127-134. http://dx.doi.org/10.1016/j.enbuild.2015.09.015

[2] Ascione, F., Bellia, L., Capozzoli A. & Minichiello, F. (2009). "Energy saving strategies in air-conditioning for museums", *Applied Thermal Engineering*, 29, pp. 676-686. doi:10.1016/j.applthermaleng.2008.03.040

[3] Chimack, M.J., Walker, Ch.E. & Franconi, E. (2001). "Determining baseline energy consumption and peak cooling loads of a 107-year-old science museum using DOE 2.1E", *Seventh International IBPSA Conference, Rio de Janeiro, Brazil, August 13-15 2001*, pp. 191-197.

[4] Zannis, G., Santamouris, M., Geros, V., Karatasou, S., Pavlou, K. & Assimakopoulos, M.N. (2006). "Energy efficiency in retrofitted and new museum buildings in Europe", *International Journal of Sustainable Energy*, 25(3-4), pp. 199-213. https://doi.org/10.1080/14786450600921645

[5] Tombazis, A.N. & Kontomichali, E.N. (2000). "Energy efficient and sustainable cultural buildings", *in Proceedings of PLEA 2000, Architecture, City, Environment, Cambridge, UK*, 2000, pp. 27-31, James & James Science Publishers.

[6] Kramer, R.P., Schellen, H.L.H. & Van Schijndel, A.W.M.J. (2016). "Impact of ASHRAE's museum climate classes on energy consumption and indoor climate fluctuations: Full scale measurements in museum Hermitage Amsterdam", *Energy and Buildings*, 130, pp. 286-294. http://dx.doi.org/10.1016/j.enbuild.2016.08.016

[7] Van Schijndel, A.W.M.J. & Schellen, H.L.H. (2018). "Mapping future energy demands for European museums", *Journal of Cultural Heritage*, 31, pp. 189-201. https://doi.org/10.1016/j.culher.2017.11.013

[8] Sharif-Askari, H. & Abu-Hijleh, B. (2018). "Review of museums indoor environment condition studies and guidelines and their impact on the museum artifacts and energy consumption", *Building and Environment*, 143, pp. 186-195. https://doi.org/10.1016/j.buildenv.2018.07.012

[9] Kramer, R.P., Maas, M.P.E., Martens, M.H.J., Van Schijndel, A.W.M. & Schellen, H.L. (2015). "Energy conservation in museums using different set-point strategies: A case study for a state-of-the-art museum using building simulation", *Applied Energy*, 158, pp. 446-458. http://dx.doi.org/10.1016/j.apenergy.2015.08.044

[10] Kompatchher, K., Seuren, S., Kramer, R., Van Schijndel, J. & Schellen, H. (2017). "Energy efficient HVAC control in historical buildings: A case study for the Amsterdam museum", *Energy Procedia*, 132, pp. 891-896. 10.1016/j.egypro.2017.09.703

[11] Kramer, R., Schellen, H. & Van Schijndel, J. (2015). "Energy impact of ASHRAE's museum climate classes: A simulation study on four museums with different quality of envelopes", *Energy Procedia*, 78, pp. 1317-1322. doi: 10.1016/j.egypro.2015.11.147

[12] Santoli, L.D., Mancini, F., Rossetti, S. & Nastasi, B. (2016). "Energy and system renovation plan for Galleria Borghese, Rome", *Energy and Buildings*, 129, pp. 549-562. http://dx.doi.org/10.1016/j.enbuild.2016.08.030

[13] Galatioto, A., Ciulla, G. & Ricciu, R. (2017). "An overview of energy retrofit actions feasibility on Italian historical buildings", *Energy*, 137, pp. 991-1000. DOI: 10.1016/j.energy.2016.12.103

[14] Mueller, H.F.O. (2013). "Energy efficient museum buildings", *Renewable Energy*, 49, pp. 232-236. doi:10.1016/j.renene.2012.01.025

[15] Cabeza, L.F., De Gracia, A. & Pisello, A.L. (2018). "Integration of renewable technologies in historical and heritage buildings: A review", *Energy and Buildings*, 177, pp. 96-111. https://doi.org/10.1016/j.enbuild.2018.07.058

[16] Moschella, A., Salemi, A., Lo Faro, A., Sanfilippo, G., Detommaso, M. & Privitera, A. (2013). "Historic buildings in Mediterranean area and solar thermal technologies: architectural integration vs preservation criteria", *Energy Procedia*, 42, pp. 416-425. doi: 10.1016/j.egypro.2013.11.042

[17] Emmi, G., Zarrella, A., Carli, M.De., Moretto, S., Galgaro, A., Cultrera, M., Tuccio, M.Di. & Bernardi, A. (2017). "Ground source heat pumps in Historical buildings: Two Italian case studies", *Energy Procedia*, 133, pp. 183-194. Doi:10.1016/j.egypro.2017.09.383

[18] Pacchiega, C. & Fausti, P. (2017). "A study on the energy performance of a ground source heat pump utilized in the refurbishment of an historical building: Comparison of different design options", *Energy Procedia*, 133, pp. 349-357. Doi:10.1016/j.egypro.2017.09.360

[19] Lucchi, E. (2018). "Simplified assessment method for environmental and energy quality in museum buildings", *Energy and Buildings*, 117, pp. 216-229. http://dx.doi.org/10.1016/j.enbuild.2016.02.037

[20] Pisello, A.L., Petrozzi, A., Castaldo, V.L. & Cotana, F. (2014). "Energy refurbishment of historical buildings with public function: Pilot case study", *Energy Procedia*, 61, pp. 660-663. doi: 10.1016/j.egypro.2014.11.937

[21] Polo Lopez, Cr. & Frontini, Fr. (2014). "Energy efficiency and renewable solar energy integration in heritage historic buildings", *Energy Procedia*, 48, pp. 1493-1502. doi: 10.1016/j.egypro.2014.02.169

[22] Rota, M., Corgnati, S.P. & Corato, L.Di. (2015). "The museum in historical buildings: Energy and systems. The project in Fondazione Musei Senesi", *Energy and Buildings*, 95, pp. 138-143. http://dx.doi.org/10.1016/j.enbuild.2014.11.008

[23] "Science museum renewable energy project". Retrieved from http://fliphtml5.com/xdom/qlee at 27/5/2019

[24] Vourdoubas, J. (2019). "Energy consumption and carbon emissions in an Academic Institution in Greece: Can it become carbon neutral?", *Studies in Engineering and Technology*, 6(1), pp. 16-23. doi:10.11114/set.v6i1.4013

[25] Vourdoubas, J. (2018). "Energy consumption and carbon emissions in Venizelio hospital in Crete, Greece: Can it be carbon neutral?", *Journal of Engineering and Architecture,* 6(1), pp. 19-27. DOI: 10.15640/jea.v6n1a2

[26] http://www.mae.com.gr/

[14] Possibilities of creating net zero carbon emission prisons in the island of Crete, Greece.

1. Introduction

The global threat of climate change requires the replacement of fossil fuels used during the last two hundred years with green benign renewable energy resources. Public buildings consume large amounts of energy while among them are included prisons and correction buildings. Current European and Greek legislation makes obligatory the energy renovation of prison buildings and their transformation towards nearly zero-energy buildings (NZEBs). Reports on the use of sustainable energies in prisons are limited so far while studies regarding the reduction of energy consumption and carbon emissions in Greek prisons are lacking. The possibility of using renewable energy technologies (RETs) for heat and power generation in prisons is investigated with reference to a prison located in Crete, Greece. Locally available renewable energies can be used resulting in minimizing or zeroing the prison's carbon footprint due to energy use. Energy refurbishment of public buildings like hospitals, schools, museums and prisons reducing their energy consumption and carbon emissions is important and necessary for improving their energy efficiency and mitigating climate change according to global, European and national targets. It is also important for achieving carbon neutrality of Europe by 2050.

2. Literature survey

2.1 Energy consumption in prisons

The energy sustainability in prisons through the implementation of the project E-SEAP financed by EU has been studied [1]. The authors mentioned that the average annual energy consumption in prisons located in seven EU countries varies between 4,000 KWh/prisoner to 16,500 KWh/prisoner while their annual carbon emissions varied between 0.6 $tnCO_2$/prisoner and 4 $tnCO_2$/prisoner. Annual energy consumption in Greek prisons has been reported at 4,000 KWh/prisoner while their annual carbon emissions at 1.2 $tnCO_2$/prisoner. It also stated that the building stock in EU prisons is old and their energy consumption is high while the possibility for energy renovation in their buildings is limited. The energy consumption profile of the existing building stock in Greece has been reported [2]. The author mentioned that total annual primary energy consumption in prisons is at 663.48 KWh/m². He also stated that the share of electricity in their total energy consumption is at 73.41%, of heating oil at 25.27%, of natural gas at 1.03% and of solar energy at 0.71%. Energy efficiency in correctional facilities in USA has been studied [3]. The author mentioned that annual primary energy intensity in prisons is at 535.2 KWh/m^2 while their on-site energy intensity is at 293.6 KWh/m^2. He also stated that prisons should ask support from state energy offices for implementation of sustainable energy projects due to high capital required and the necessity for qualified energy expertise. A report on four existing prisons located in Crete, Greece has been published [4]. The report mentioned that the capacity of the existing prisons in Crete is as follows: Chania prison - 480 offenders, Agia prison - 178 offenders, Nea Alikarnasos prison - 210 offenders and Neapoli prison - 45 offenders. Various definitions of net zero energy buildings have been reported [5]. The authors mentioned four definitions as follows: net zero site energy, net zero source energy, net zero energy costs and net zero energy emissions. A net zero energy emissions building, according to them, produces at least as much emissions-free renewable energy as it uses from emissions-producing energy sources. An overview of life cycle energy use in buildings has been presented [6]. The authors analyzed results of 73 cases from 13 countries including residential and office buildings. They found that operating energy corresponds at 80%-90% of the life cycle energy use and the rest corresponds to embodied energy. They estimated the annual life cycle primary energy requirements in residential buildings at 150-400 KWh/m^2 while in offices at 250-550 KWh/m^2. A study for minimizing the life cycle energy in buildings has been published [7]. The authors stated that in mild climates embodied energy in buildings represents up to 25% of the total life cycle energy use. In the future when operating energy in buildings will be reduced due to the construction of nearly-zero energy buildings the ratio of embodied energy to total life cycle energy is expected to increase. A report on prisons' sustainability has been presented [8]. The authors mentioned that reduction of energy consumption and carbon emissions are required in prisons in order to increase their sustainability. They set targets regarding the maximum daily energy consumption in prisons at 45-60 KWh/offender. A study of the existing correctional system focusing on USA has been implemented [9]. The authors stated that the notion of "green prisons" as well as environmental and energy regulations should be related with the human treatment of prisoners. They expressed their skepticism on the priority of "greening the correctional buildings" over the issue of "social justice".

2.2 Use of sustainable energies in prisons

A study regarding use of a solar thermal system for hot water production in a prison located in Philadelphia, USA has been realized [10]. The author mentioned that the prison hosts 2,500 prisoners. Installation of 248 solar panels on the rooftop of their buildings could reduce their energy consumption by 20-25%. The cost of fuel used for hot water production as well as their carbon emissions could be also substantially reduced. An evaluation of biogas sanitation systems in Nepalese prisons has been made [11]. The authors mentioned that biogas digesters have been installed in three Nepalese prisons fermenting human wastes mixed with kitchen wastes. The digester's daily productivity was 62 NL/person while the methane concentration was at a volume of 57-78%. The produced biogas was used as cooking fuel reducing the fuel's cost in the prisons. An investigation of the possibility of installing a biogas plant processing the organic wastes in Tihar prison in India has been implemented [12]. The author estimated the organic waste production at 750 kg/day and the biogas potential at 67.5 NM^3/day. He stated that payback period of the investment was attractive at 2.16 years. A report on the Santa Rita "green prison" with capacity of 4,000 offenders located nearby San Francisco, USA has been published [13]. The authors mentioned that the prison had various renewable energy technology installations including a 1.2 MW solar-PV system, a 1 MW fuel cell with heat recovery while installation of 2 MW of NaS batteries was planned. They also stated that the installed distributed energy generation technologies resulted in minimization of jail's energy bill as well as in its carbon emissions. A study on the energy consumption and carbon emissions in Venizelio hospital located in Crete, Greece has been realized [14]. The author mentioned that its annual energy consumption was at 280 KWh/m^2 while the annual carbon emissions due to energy use were at 168 kgCO_2/m^2. He also stated that use of various renewable energy technologies including solar thermal systems, solar-PV systems, solid biomass burning systems and ground source heat pumps could zero its net annual carbon emissions into the atmosphere. Creation of net zero carbon emission hotels in Mediterranean region has been studied by Vourdoubas, 2018 [15]. The author mentioned that since solar energy is abundant in Mediterranean region its use in hotels is suggested. Solar thermal technology for hot water production and solar-PV technology for electricity generation combined with high efficiency heat pumps could cover all their energy requirements resulting in carbon-neutral hotels. The "greening" of correction buildings has been investigated [16]. The authors mentioned that prisons consist of the most energy-intensive buildings and offer great opportunity for energy savings. They also stated that measures towards improvement of energy behavior in prisons require participation of all stakeholders including the offenders. A study regarding the existing four rural prisons in Greece has been made [17]. Among them one is located in the village of Agia in Western Crete. Rural prisons have enough space for installation of the necessary sustainable energy systems generating green heat and electricity while biomass produced in their fields could be used for production of various energy products.

2.3 Carbon sequestration from tree plantations

An investigation of the effect of weeds on net CO_2 exchange in an olive tree orchard in Spain has been implemented [18]. The authors compared the net CO_2 exchange in two olive orchards, one with weed cover and one without. They stated that the net annual CO_2 exchange in the olive orchard covered with weeds was double than in the other at 0.513 $kgCO_2/m^2$. An estimation of the carbon footprint in an olive tree grove located in central Italy over a period of 11 years has been made [19]. The authors compared carbon sequestration and carbon emissions in the olive tree plantation over this period. They estimated that carbon sequestration exceeded carbon emissions in the fourth year while the net annual carbon sequestration over this period was 3.45 $kgCO_2/m^2$. A study on CO_2 sequestration from olive groves in Southern Spain has been implemented [20]. The authors estimated the net carbon balance in various intensive and conventional olive tree plantations. They mentioned that the annual atmospheric CO_2 removal rate in these olive groves was varying between 0.75 $kgCO_2/m^2$ and 1.50 $kgCO_2/m^2$. Research on carbon sequestration potential of selected tree species in India using a dynamic growth model has been realized [21]. The authors stated that the net annual carbon sequestration rate for fast growing species like poplar was at 2.93 $kgCO_2/m^2$ while for Eucalyptus at 2.2 $kgCO_2/m^2$. An investigation on carbon storage in Eucalyptus plantations in Southern China has been made [22]. The authors mentioned that Eucalyptus plantations stored carbon in both the above- and underground fractions while the annual carbon sequestration rate over an eight-year period was at 3.23 $kgCO_2/m^2$. They also mentioned that Eucalyptus may deplete fast soil nutrients while they suggested the development of mixed forest communities for improving ecosystem's sustainability. Published data regarding energy consumption and carbon emissions in various prisons are presented in Table 1. Published data regarding carbon removal from tree plantations are presented in Table 2.

Table 1. Energy consumption and carbon emissions in prisons and correction buildings

Source	Annual energy consumption	Annual carbon emissions	Share of electricity in total energy use
Christoforidis et al, 2014	4,000-16,500 KWh/prisoner	0.6-4 tnCO$_2$/prisoner	
Christoforidis et al, 2014 (Greek prisons)	4,000 KWh/prisoner	1.2 tnCO$_2$/prisoner	
Petroliagki, 2018	663.48 KWh/m^2 (primary energy consumption)		73.41%
Lunn, 2014	293.6 KWh/m^2 (on-site energy consumption) 535.2 KWh/m^2 (primary energy consumption)		
LeRoy et al, 2012	14,600-21,900 KWh/prisoner		

Table 2. Atmospheric carbon removal rate in tree plantations

Author/date	Tree plantation	Annual CO$_2$ removal rate (kgCO$_2$/m^2)
Chamizo et al, 2017	Olive	0.51
Proietti et al, 2014	Olive	3.45
Lopez-Bellido et al, 2016	Olive	0.75-1.50
Kaul et al, 2010	Poplar	2.93
Kaul et al, 2010	Eucalyptus	2.2
Du et al, 2015	Eucalyptus	3.23

The aims of the current work are:
a) The investigation of the possibility of using renewable energy technologies for covering all the energy needs in prisons in Crete, and

b) The estimation of the size and cost of the required renewable energy systems as well as the area of the required tree plantations for carbon sequestration for achieving a net zero carbon emissions prison located in Crete, Greece

The methodology followed includes:

a) the use of existing data regarding the energy consumption in prisons and the share of electricity in their energy mix, b) investigation of various renewable energy technologies which could be used in Cretan prisons, c) implementation of a case study regarding zeroing net carbon emissions in a Cretan prison, estimating the size of the proposed renewable energy systems and the area of the tree plantation required for carbon sequestration, and d) presentation of existing opportunities and barriers for greening existing prisons in Crete. Uncertainties in the current study are related with limited published data regarding energy consumption and carbon emissions in Greek prisons and correction buildings.

3. Energy consumption in prisons

3.1 Operating energy

Prisons consume energy in various sectors including space heating and cooling, hot water production, lighting and operation of various electric machinery and devices. Specific energy consumption in prisons varies and depends on the type of construction of the building, the local climate and the type of prison. Although, to my best knowledge, there are no published data regarding life cycle energy consumption in prison buildings, it will be assumed that they follow the same pattern like in other buildings and their operating energy has a share at approximately 85% of their life cycle energy consumption. Annual energy consumption in Greek prisons has been reported at 4,000 KWh/prisoner while their annual carbon emissions at 1.2 $tnCO_2$/prisoner. It should be noted that the reported energy consumption in European and USA prisons is considerably higher than the values reported in Greek prisons.

3.2 Embodied energy

The energy consumed during prison's construction, renovations and demolition is called embodied energy and it is assumed that it corresponds at 15% of its life cycle energy use. Embodied energy in Greek prisons is estimated at 706 KWh/prisoner per year. The fraction of embodied energy to life cycle energy use in conventional buildings is relatively low. However, if prison buildings are transformed to nearly-zero energy buildings and their operating energy is reduced then the fraction of embodied energy to life cycle energy use would be increased. Taking into account that the operating energy use is 4,000 KWh/prisoner and the embodied energy use is 706 KWh/prisoner it is concluded that the annual life cycle energy use in Greek prisons is 4,706 KWh/prisoner.

4. Possibilities of using renewable energy technologies in Cretan prisons

Various renewable energy technologies which are mature, reliable and cost-effective could be used in Cretan prisons for covering their energy requirements in heat, cooling and electricity. These include:
a) Solar thermal energy for hot water production,
b) Solar-PV energy for electricity generation,
c) Solid biomass burning for heat generation, and
d) High efficiency heat pumps for space heating and cooling.

Solar energy is broadly used in Mediterranean region for hot water production in buildings with simple thermosiphonic systems and flat plate collectors. Solar-PV energy is used for electricity generation in grid-connected buildings particularly during the last decade when the cost of solar panels has been substantially decreased and generation of solar electricity has become profitable. Net-metering regulations allow storage of electricity into the grid when it is not needed in the building. Solid biomass can be used for heat generation by burning either in stoves or in open fires. Finally, high efficiency heat pumps are broadly used for heat and cooling generation. Renewable energy technologies which can be used in Cretan prisons are presented in Table 3.

Table 3. Renewable energy technologies which can be used for energy generation in Cretan prisons

Energy source	Technology	Space heating	Hot water production	Space cooling	Electricity generation
Solar energy	Flat plate collectors - thermosiphonic systems		Yes		
Solar energy	Photovoltaic panels				Yes
Solid biomass	Burning	Yes	Yes		

| Ambient heat | High efficiency heat pumps | Yes | Yes | Yes | |

5. Requirements for net zero carbon emissions prisons

Prisons, like other buildings, are obliged to reduce their energy consumption and replace the use of fossil fuels with renewable energies. Many sustainable energy technologies can be used which are mature, reliable and cost-effective. A carbon neutral prison due to operating energy use should:

a) Reduce its energy consumption using energy-saving techniques,
b) Replace all the fossil fuels used with renewable energies like solar energy or biomass,
c) Generate the same amount of grid electricity used annually with green electricity generated preferably with solar-PV panels according to the net-metering regulations. The solar panels can be installed either on-site or off-site of its premises,
d) Alternatively, carbon emissions due to fossil fuels use could be offset by creation of tree plantations which sequestrate atmospheric carbon equal to the emitted quantities.

If a prison complies with the abovementioned requirements it could zero its net annual carbon emissions due to operating energy use.

6. A case study of a net zero carbon emissions prison located in Crete, Greece

For achieving a net zero carbon emissions prison in Crete, Greece, two scenarios have been examined. In the first scenario solar energy systems can generate annually all the required energy in the prison, eliminating the need for fossil fuels use or electricity derived from them. In the second scenario, half of the required energy in the prison will be generated by solar energy and the rest by fossil fuels. Additionally, a tree plantation will be created which will sequestrate annually the same amount of carbon that is emitted by the fossil fuels use in the prison.

6.1 Sizing the solar energy systems generating all energy requirements

The possibility of zeroing net carbon emissions in Agia's prison with a capacity of 178 offenders located in western Crete is investigated. In the first scenario solar energy systems are used for generation of all the required energy in the prison. In order to size the renewable energy systems, the following assumptions are made:

a) The energy and the fuels currently used for hot water production in the prison are electricity and diesel oil. The specific annual energy consumption is 4,000 KWh/offender while 25% of its energy consumption is used for hot water production and the remaining 75% for lighting and for powering electric devices including the equipment used for space heating and cooling, which use 40% of the overall electricity consumption.

b) The sustainable energy systems that will be used in the prison include: A solar thermal system for hot water production, a solar-PV system for electricity generation, and a high efficiency heat pump for air conditioning. There is enough space availability for installing the solar energy systems required, since Agia prison is an "rural prison".

c) The above-mentioned sustainable energy systems can generate annually all the energy required when the prison is full. The prison is grid-connected and the generated solar electricity, when it is not needed, will be stored in the grid according to net-metering regulations.

d) The solar thermal system with flat plate collectors will produce 80% of hot water requirements while the remaining 20% will be produced with electricity.

e) Energy renovation in the prison will result in 20% decrease in its overall energy consumption which is distributed equally in all sectors.

f) Annual electricity generation by a solar-PV system in Crete equals at 1,500 KWh/KW$_p$, annual heat generation by a solar thermal system equals at 500 KWh$_{th}$/m^2 of flat plate collectors while the capital cost of a solar-PV system is at 1,200 €/KW$_p$, and the capital cost of a solar thermal system at 300 €/m^2 of collectors. The size of the solar energy systems generating all the required energy in Agia prison are presented in Table 4.

Table 4. Size of solar energy systems for zeroing net annual carbon emissions due to operating energy use in Agia prison in Crete, Greece

Parameter	Value
Initial annual energy consumption	712,000 KWh
Annual energy consumption after energy refurbishment	569,600 KWh
Annual electricity use in all sectors except in hot water production	427,200 KWh$_{el}$
Annual requirements for hot water production	142,400 KWh$_{th}$

Annual electricity requirements for hot water production	28,480 KWh$_{el}$
Annual heat generation by the solar thermal system	113,920 KWh$_{th}$
Total annual electricity requirements	455,680 KWh$_{el}$
Nominal power of a solar-PV system	303.8 KW$_p$
Size of a solar thermal system - area of flat plate collectors	227.8 M^2
Capital cost of the solar-PV system	364,560 €
Capital cost of the solar thermal system	68,340 €
Capital cost of both solar energy systems	432,900 €
Overall capital cost per offender	2,432 €
Annual carbon emissions due to heating oil use[1]	38.45 tnCO$_2$
Annual carbon emissions due to grid electricity use[2]	256.32 tnCO$_2$
Annual total carbon emissions due to conventional fuel use	294.77 tnCO$_2$
Overall annual carbon emissions per offender due to conventional fuel use	1.66 tnCO$_2$

[1]*Carbon emission: heating oil, 0.27 kgCO$_2$/KWh,* [2]*Carbon emissions, grid electricity, 0.6 kgCO$_2$/KWh*

6.2 Sizing the solar energy systems and the area of the tree plantation for zeroing carbon emissions

In the second scenario it is assumed that the solar energy systems will generate half of the required energy in the prison while the rest will be produced by fossil fuels. Tree plantations either with olive trees or by Eucalyptus species will be created for sequestration of CO$_2$ emissions. The following estimations are based in these assumptions. The size of solar energy systems and the area of the required tree plantations are presented in Table 5.

Table 5. Size of the solar energy systems which will generate half of the required energy while CO$_2$ emitted by fossil fuels will be offset by a tree plantation

Parameter	Value
Annual electricity use in all sectors except in hot water production	213,600 KWh$_{el}$
Annual requirements in hot water production	71,200 KWh$_{th}$

Annual electricity requirements in hot water production	14,240 KWh$_{el}$
Annual heat generation by the solar thermal system	56,960 KWh$_{th}$
Total annual electricity requirements	227,840 KWh$_{el}$
Size of a solar-PV system	151.9 KW$_p$
Size of a solar thermal system - area of flat plate collectors	113.9 M^2
Capital cost of the solar-PV system	182,280 €
Capital cost of the solar thermal system	34,170 €
Capital cost of both solar energy systems	216,450 €
Overall capital cost per offender	1,216 €
Annual carbon emissions due to heating oil use [1]	19.2 tnCO$_2$
Annual carbon emissions due to grid electricity use [2]	128.1 tnCO$_2$
Annual total carbon emissions due to conventional fuel use	147.3 tnCO$_2$
Overall annual carbon emissions per offender due to conventional fuel use	0.83 tnCO$_2$
Area of an olive tree plantation sequestrating all carbon emissions [3]	14.74 ha
Area of a Eucalyptus tree plantation sequestrating all carbon emissions [4]	5.9 ha

[1]*Carbon emissions: heating oil, 0.27 kgCO$_2$/KWh,* [2]*Carbon emissions: grid electricity, 0.6 kgCO$_2$/KWh,* [3]*Annual carbon sequestration by olive trees: 1 kgCO$_2$/m^2,* [4]*Annual carbon sequestration by Eucalyptus trees: 2.5 kgCO$_2$/m^2*

7. Opportunities and barriers in "greening" Cretan prisons

Existing opportunities for creation of net zero carbon emissions prisons in Crete are based on the fact that:

a) European policy promotes creation of nearly zero energy buildings which also have nearly zero or zero carbon emissions prioritizing public buildings,

b) Net-metering regulations allow the generation of solar green electricity in prisons and its storage into the grid when it is not needed,

c) Various solar energy technologies for heat and electricity generation can assist in creation of net zero carbon emissions prisons. They are mature, reliable and cost-effective while solar energy irradiance is high in Crete, and

d) There are financial instruments like third-party financing through energy service companies which can support the implementation of energy renovation projects in prisons.

There are various internal weaknesses in Greek prisons negatively influencing their "greening" including:

a) Lack of financial resources due to current economic crisis in Greece,
b) Lack of technical expertise and staff required for supervision and maintenance of the sustainable energy systems installed in the prisons, and
c) Lack of organizational flexibility and required human capacity in them.

8. Discussion

Estimations regarding the use of solar energy systems for energy generation in Agia's prison in Crete, Greece have been implemented. Additionally, the area of a tree plantation offsetting carbon emissions due to fossil fuels use in the prison has been calculated. These estimations regarding the rural prison in Agia, Crete indicated that creation of net zero carbon emissions correctional buildings due to energy use are feasible in Crete, Greece. Their "greening" could be achieved with combined use of solar energy technologies which are currently reliable, cost-effective and broadly used. Additionally, carbon emissions offsetting with creation of tree plantations sequestrating atmospheric carbon can be considered. Results can be used for energy renovation of existing prison buildings in Crete in compliance with the current Greek and EU legislation. Prisoners could be engaged in operation and maintenance of the installed solar energy systems and probably in planting and irrigation of tree plantations. They can be also trained in sustainable energy technologies through seminars, organized inside the prison, in order to increase their knowledge and skills in these benign energy technologies. The acquired knowledge and skills can assist them in their future employment. Further work should be focused on accurate estimation of energy consumption in Cretan prisons and its distribution in each sector. A cost-benefit analysis of investments in sustainable energy systems achieving net zero carbon emissions due to operating energy use in them can be useful in their future energy refurbishment.

9. Conclusions

The possibility of zeroing net annual carbon emissions due to operating energy use in Cretan prisons with reference the agricultural prison in Agia, Crete, Greece has been examined. The size and cost of renewable energy systems achieving that have been estimated. Finally, the area of the required tree plantation offsetting carbon emissions due to fossil fuels use in the prison has been calculated. It has been indicated that using mature, reliable and cost-effective solar energy technologies for heat and electricity generation the annual net carbon emissions due to energy use in Agia's prison can be zeroed. Additionally, tree plantations can be created for offsetting any CO_2 emitted into the atmosphere due to fossil fuels use. Agia's prison can host 178 offenders, its annual operational energy consumption is at 4,000 KWh/prisoner, while its annual CO_2 emissions due to fossil fuels use is at 1.66 $kgCO_2$/prisoner. The surface of flat plate solar collectors in the solar thermal system has been estimated at 113.9 M^2 to 227.8 M^2 while its cost is estimated at 34,170 € to 68,340 €. The nominal power of the solar-PV system has been estimated at 151.9 KW_p to 303.8 KW_p while its cost at 182,280 € to 364,560 €. The area of the tree plantation sequestrating 50% of the annual CO_2 emissions due to fossil fuels use in the prison has been estimated at 14.74 ha for olive trees and at 5.9 ha for Eucalyptus trees. The results indicated that energy refurbishment in Agia's rural prison in Crete, Greece in order to zero its annual net carbon emissions due to energy use is feasible using existing commercial green solar energy technologies.

References

[1] Christoforidis, G.C., Papagiannis, G.K., Brain, M. & Puksec, T. (2014). "Establishing an assessment framework for energy sustainability in prisons: The E-SEAP project", in the *14th International Conference on Environment and Electrical Engineering (EEEIC)*, May 2014. DOI: 10.1109/EEEIC.2014.6835861

[2] Petroliagki, M. (2018). "Energy consumption profile of the existing building stock in Greece", Greek Ministry of Environment and Energy. Retrieved at 24/3/2020 from http://bpes.ypeka.gr/wp content/uploads/2018_03_03.PRESENTATION_ENERGY_POVERTY-%CE%94%CE%99%CE%9F%CE%A1%CE%98%CE%A9%CE%9C%CE%95%CE%9D%CE%9F.pdf

[3] Lunn, M. (2014). "Energy efficiency in correctional facilities and opportunities for State energy office engagement", U.S. Department of Energy, Energy efficiency and Renewable Energy, 23/1/2014. Retrieved at 24/3/2020 from https://www.energy.gov/sites/prod/files/2014/05/f15/tap_corrections_presentation.pdf

[4] "Correction buildings in Crete", (2017), (In Greek). Retrieved at 24/3/2020 from https://www.cretapost.gr/347105/aktinografia-ton-sofronistikon-katastimaton-tis-kritis/

[5] Torcellini, P., Pless, S., Deru, M. & Crawley, D. (2006). "Zero Energy Buildings: A Critical Look at the Definition", National Laboratory of the US Department of Energy. *ACEEE Summer Study*, 14-18 August 2006, Pacific Grove, California.

[6] Ramesh, T., Prakash, R. & Shukla, K.K. (2010). "Life Cycle Energy Analysis of Buildings: An Overview", *Energy and Buildings*, 42, pp. 1592-1600. https://doi.org/10.1016/j.enbuild.2010.05.007

[7] Karimpour, M., Belusko, M., Xing, K. & Bruno, F. (2014). "Minimizing the Life Cycle Energy of Buildings: Review and Analysis", *Building and Environment*, 73, pp. 106-114. https://doi.org/10.1016/j.buildenv.2013.11.019

[8] LeRoy, C.J., Bush, K., Trivett, J. & Gallagher, B. (2012). "The sustainability in prisons project. An overview, 2004-2012". Retrieved at 27/3/2020 from http://sustainabilityinprisons.org/wp-content/uploads/2016/02/Overview-cover-text-reduced-size.pdf

[9] Moran, D. & Jewkes, Y. (2014). "Green prisons rethinking the "sustainability" of the Carceral estate", *Geographica Helvetica*, 69, pp. 345-353. doi:10.5194/gh-69-345-2014

[10] Eskind, R. (2009). "Solar hot water system installed in Philadelphia prison", *Corrections Today*, June 2009, pp. 68-72.

[11] Lohri, C., Vogeli, Y., Oppliger, A. Mardini, R. Giusti, A. & Zurbrugg, C. (2010). "Evaluation of biogas sanitation systems in Nepalese prisons, In Decentralized wastewater treatment solutions in developing countries", *Conference and Exhibition, Surabaya, Indonesia*, 23-26 March 2010. Retrieved at 24/3/2020 from https://www.researchgate.net/publication/229010430_Evaluation_of_Biogas_Sanitation_Systems_in_Nepalese_Prisons

[12] Krishnan, A. (2015). "A study of challenges, opportunities and feasibilities of installation of biogas plants across two institutions – A case study of a jail and a zoo". Retrieved at 26/3/2020 from https://www.academia.edu/34542424/A_STUDY_OF_CHALLENGES_OPPORTUNITIES_AND_FEASIBILITIES_OF_INSTALLATION_OF_BIOGAS_PLANTS_ACROSS_TWO_INSTITUTIONS_-A_CASE_STUDY_OF_A_JAIL_AND_A_ZOO

[13] Marnay, Ch., DeForest, N., Stadler, M., Donadee, J., Dierckxsens, C., Mendes, G., Lai, J. & Ferreira Cardose, G. (2011). "A green prison: Santa Rita jail creeps towards zero net energy", in *ECEEE summer study*, 6-11 June 2011, Belambra Presquile de Giens, France. https://escholarship.org/uc/item/9xz7c3nz

[14] Vourdoubas, J. (2018). "Energy consumption and carbon emissions in Venizelio hospital in Crete, Greece: Can it become carbon neutral?", *Journal of Engineering and Architecture*, 6(1), pp. 19-27. DOI: 10.15640/jea.v6n1a2

[15] Vourdoubas, J. (2018). "Hotels with net zero carbon emissions in the Mediterranean region: Are they feasible?", *Journal of Tourism and Hospitality Management*, 6(2), pp. 72-79. DOI: 10.15640/jthm.v6n2a6

[16] Cross, J.E., O'Conner Shelley, T. & Mayer, A.P. (2017). "Putting the green into corrections: Improving energy conservation, building function, safety and occupant well-being in an American correction facility", *Energy Research & Social Science*. . http://dx.doi.org/10.1016/j.erss.2017.06.020

[17] Schizas, P. (2013). "Rural prisons in Greece", Ph.D. thesis, Panteion University of Social and Political Sciences, Athens, Greece, (in Greek). Retrieved at 28/3/2020 from http://thesis.ekt.gr/thesisBookReader/id/28700#page/10/mode/2up

[18] Chamizo, S., Serrano-Ortiz, P., Lopez-Ballesteros, A., Sanchez-Canete, E.P., Vicente-Vicente, J.L. & Kowalski, A.S. (2017). "Net ecosystem CO_2 exchange in an irrigated olive orchard of SE Spain: Influence of weed cover", *Agriculture, Ecosystems and Environment*, 239, pp. 51-64. http://dx.doi.org/10.1016/j.agee.2017.01.016

[19] Proietti, S., Sdringola, P., Desideri, U., Zepparelli, F., Brunori, A., Ilarioni, L., Nasini, L., Regni, L. & Proietti, P. (2014). "Carbon footprint of an olive tree grove", *Applied Energy*, 127, pp. 115-124. http://dx.doi.org/10.1016/j.apenergy.2014.04.019

[20] Lopez-Bellido, P.J., Lopez-Bellido, L., Fernandez-Garcia, P., Munoz-Romero, V. & Lopez-Bellido, F.J. (2016). "Assessment of carbon sequestration and the carbon footprint in olive groves in Southern Spain", *Carbon Management*, 7(3-4), pp. 161-170. https://doi.org/10.1080/17583004.2016.1213126

[21] Kaul, M., Mohren, G.M.J. & Dadhwal, V.K. (2010). "Carbon storage and sequestration potential of selected tree species in India", *Mitigation and Adaptation Strategies for Global Change*, 15(5), pp. 489-510. DOI :10.1007/s11027-010-9230-5

[22] Du, H., Zeng, F., Peng, W., Wang, K., Zhang, H., Liu, L. & Song, T. (2015). "Carbon storage in a Eucalyptus plantation Chronosequence in Southern China", *Forests*, 6, pp. 1763-1778. doi:10.3390/f6061763

[15] Papers included in the book

Part 1

[1] Vourdoubas, J. (2016). "Creation of zero CO_2 emissions residential buildings due to energy use: A case study in Crete-Greece", *JOURNAL OF CIVIL ENGINEERING AND ARCHITECTURE RESEARCH*, 3(2), pp. 1251-1259.

[2] Vourdoubas, J. (2017). "Realization of a small residential building with zero CO_2 emissions due to energy use in Crete, Greece", *STUDIES IN ENGINEERING AND TECHNOLOGY*, 4(1), pp. 112-120.

[3] Vourdoubas, J. (2017). "Creation of zero CO_2 emission residential buildings due to operating and embodied energy use on the island of Crete, Greece", *OPEN JOURNAL OF ENERGY EFFICIENCY*, 6(4), pp. 141-154.

[4] Vourdoubas, J. (2018). "Use of renewable energies for the creation of net zero carbon emission residential buildings in northern Greece", *OPEN JOURNAL OF ENERGY EFFICIENCY*, 7(3), pp. 75-87.

[5] Vourdoubas, J. (2018). "Review of sustainable energy technologies used in buildings in the Mediterranean basin", *JOURNAL OF BUILDINGS AND SUSTAINABILITY*, 1(2), pp. 1-11.

[6] Vourdoubas, J. (2020). "Creation of net zero carbon emissions residential buildings due to energy use in the Mediterranean region: Are they feasible?", *CIVIL ENGINEERING RESEARCH JOURNAL*, 10(1), 555777

Part 2

[7] Vourdoubas, J. (2015). "Creation of zero CO_2 emissions hospitals due to energy use. A case study in Crete-Greece", *JOURNAL OF ENGINEERING AND ARHITECTURE*, 3(2), pp. 79-86.

[8] Vourdoubas, J. (2016). "Creation of zero CO_2 emissions school buildings due to energy use in Crete-Greece", *OPEN JOURNAL OF ENERGY EFFICIENCY*, 5(1), pp. 12-18.

[9] Vourdoubas, J. (2016). "Possibilities of creating swimming pools with zero CO_2 emissions due to energy use in them", *ASIAN JOURNAL OF ENERGY TRANSFORMATION AND CONSERVATION*, 3(1), pp. 1-9.

[10] Vourdoubas, J. (2017). "Creation of zero CO_2 emission office buildings due to energy use: A case study in Crete, Greece", *INTERNATIONAL JOURNAL OF MULTIDISCIPLINARY RESEARCH AND DEVELOPMENT*, 4(2), pp. 165-170.

[11] Vourdoubas, J. (2018). "Energy consumption and carbon emissions in Venizelio hospital in Crete, Greece: can it be carbon neutral?", *JOURNAL OF ENGINEERING AND ARHITECTURE*, 6(1), pp. 19-27.

[12] Vourdoubas, J. (2019). "Energy consumption and carbon emissions in an Academic Institution in Greece: Can it become carbon neutral?", *STUDIES IN ENGINEERING AND TECHNOLOGY*, 6(1), pp. 16-23.

[13] Vourdoubas, J. (2019). "Possibilities of creating net zero carbon emissions cultural buildings: A case study of the museum at Eleftherna, Crete, Greece", *AMERICAN SCIENTIFIC RESEARCH JOURNAL FOR ENGINEERING, TECHNOLOGY AND SCIENCES*, 56(1), pp. 207-217.

[14] Vourdoubas, J. (2020). "Possibilities of creating net zero carbon emissions prisons in the island of Crete, Greece", *OPEN JOURNAL OF ENERGY EFFICIENCY*, 9(2), pp. 81-93.

Publisher: Eliva Press SRL

Email: info@elivapress.com

All rights reserved

Eliva Press is an independent publishing house established for the publication and dissemination of academic works all over the world. Company provides high quality and professional service for all of our authors.

Our Services:
Free of charge, open-minded, eco-friendly, innovational.

-Free standard publishing services (manuscript review, step-by-step book preparation, publication, distribution, and marketing).
-No financial risk. The author is not obliged to pay any hidden fees for publication.
-Editors. Dedicated editors will assist step by step through the projects.
-Money paid to the author for every book sold. Up to 50% royalties guaranteed.
-ISBN (International Standard Book Number). We assign a unique ISBN to every Eliva Press book.
-Digital archive storage. Books will be available online for a long time. We don't need to have a stock of our titles. No unsold copies. Eliva Press uses environment friendly print on demand technology that limits the needs of publishing business. We care about environment and share these principles with our customers.
-Cover design. Cover art is designed by a professional designer.
-Worldwide distribution. We continue expanding our distribution channels to make sure that all readers have access to our books.

www.elivapress.com

Made in the USA
Middletown, DE
24 January 2021